SCRUBBING THE SKY

SCRUBBING THE SKY

INSIDE
THE RACE TO
COOL THE
PLANET

PAUL McKENDRICK

Figure.1
Vancouver / Toronto / Berkeley

22 23 24 25 26   5 4 3 2 1

Cataloguing data is available from Library and Archives Canada
ISBN 978-1-77327-208-5 (hbk.)
ISBN 978-1-77327-209-2 (ebook)
ISBN 978-1-77327-210-8 (pdf)

Design by Teresa Bubela
Author photograph by Eva Urbanska

Editing by Steve Cameron
Copy editing by David Marsh
Proofreading by Melissa Churchill
Indexing by Emily LeGrand

Front cover image by agsandrew/stock.adobe.com

Printed and bound in China by C&C Offset Printing Co.
Distributed internationally by Publishers Group West

Figure 1 Publishing Inc.
Vancouver BC Canada
www.figure1publishing.com

Figure 1 Publishing works in the traditional, unceded territory of the xʷməθkʷəy̓əm (Musqueam), Sḵwx̱wú7mesh (Squamish), and səlilwətaɬ (Tsleil-Waututh) peoples.

MIX
Paper | Supporting responsible forestry
FSC
www.fsc.org
FSC® C008047

CONTENTS

INTRODUCTION

THE FIRST TIME I SPOKE with David Keith about direct air capture, we were huddled around a meager fire trying to keep warm. We were among a group of men, brought together by Ed Whittingham, a mutual friend, seeking to enjoy some non-household company nearly a year into the COVID-19 pandemic; it was perhaps the third or fourth time we had convened in someone's yard. Having followed climate science, I was aware that David was a Harvard professor who was well known in the field, but I was pleasantly surprised to learn that he lived in the same small town as me and that, because he was teaching over Zoom at the time, I would have the opportunity to chat with him over a beer. I was also aware that he had founded one of the world's leading direct air capture companies and, moving on from the pleasantries of previous fires, that is where I directed the conversation.

To me at the time, direct air capture, a technology that scrubs carbon dioxide from the sky, was a missing ingredient in the ongoing development of climate policy. Unlike the byzantine policies being adopted by governments that attempt to appease as many differing factions as possible, and often struggle to

1

be effective as a result, direct air capture seemingly provided something simpler: a backstop for an ultimatum to stop emitting greenhouse gases. If emitters are unable or unwilling to find alternatives (by some predetermined point in time), then why not require them to remove an offsetting amount of carbon dioxide from the atmosphere? With sufficient advance warning that such a decree was coming, capital could mobilize around bringing down the cost of complying in time. David's company, I argued, was well positioned to play an integral role as part of that backstop. But he didn't share my view—or at least that it might be as straightforward as I contemplated. As I made my way home, trying to steer my bike with frozen hands while replaying the conversation in my head, I sensed there was a story to be told.

While David would like to see my fireside fantasy become a reality, he is also circumspect about the political, economic, social and environmental realities of commercializing a new technology to take on such a momentous role. Predicting when that could be theoretically possible relies on many hypothetical scenarios that David has a unique perspective on, working as he does at the intersection of climate science, energy technology and public policy. Each time I spoke with him, I gained a better appreciation for his world, one in which he relentlessly searches for answers to how we might keep the planet as hospitable as possible while navigating political hurdles and seeking to understand and minimize the risk of unintended consequences.

BACK IN 2012, ON A SNOWY DAY at the University of Calgary, developers of direct air capture technology gathered at David's behest to discuss the nascent field—and debate its prospects. A report had been published the year prior by the American Physical Society that questioned the viability of direct air capture, which entails using chemicals to selectively bind with the carbon dioxide in air and then separating the captured carbon dioxide to either be stored underground or used to make something. At the time, the odds of direct air capture becoming a cost-effective option were generally viewed as dismal—at least by the limited number of individuals who were aware of it and predisposed to hazarding a guess—owing to the diluteness of its concentration in the atmosphere. Removing carbon dioxide from air is akin to capturing the equivalent volume of gas that would occupy the interior of an average-sized hot tub after that gas is dispersed in a volume of air the size of an Olympic pool. The report pegged the cost of doing so at around $600 (USD) per metric ton. "Significant uncertainties in the process parameters result in a wide, asymmetric range associated with this estimate, with higher values being more likely than lower ones," stated the report. "Thus, direct air capture is not currently an economically viable approach to mitigating climate change." At the meeting, members of the panel that produced the prohibitive cost estimate were persuaded that they might have overestimated costs slightly. But, they asked, "Would it really make a difference?"

A British science writer named Oliver Morton was also at the meeting. Rather than being dismayed by the seemingly intractable obstacles that stood in the way of commercializing the technology, he took heed of the messages delivered by those advocating on its behalf. Four companies developing the technology were represented at the meeting, and three of those representatives were physicists (David being one of them). "They're the sort of people who impress and inspire students by showing the near-inexhaustible ability of physics to provide answers, and by encouraging them to ask questions to which the answers are truly interesting," wrote Morton. "[They] make knowledge—both theirs and, once you learn from them, yours—feel like power." The three had formed companies tasked with making direct air capture a commercial reality in their spare time outside the classroom. After listening to them, Morton left feeling bolstered by a sense that, in time, direct air capture might make a difference.

"Knowing that helps put the present into context and lets you imagine the future more fully."

Today, a decade after the meeting in Calgary, many things have changed for the technology. An important one is that much of the world is more committed to keeping carbon in the ground, leaving less reason for direct air capture, and other possible climate interventions, to be viewed as diversionary pursuits that risk diminishing the necessary appetite for the hard work of decarbonizing. At the same time, the need for carbon removal options is even greater, as the window has effectively closed on reaching climate targets without them. The Intergovernmental Panel on Climate Change, which released its sixth assessment report in 2022, states that carbon removal is now essential if warming is to ever be limited to 1.5 degrees Celsius or less. The consequential roles assigned to carbon removal are to offset any emissions at first, then more difficult-to-abate sources of emissions as the world decarbonizes and, eventually, historical emissions, allowing us to veer the planet off a warming trajectory and onto one that may get cooler over time.

This book tells the story of a group of scientists, philanthropists, investors and advocates who have resisted strong headwinds to provide us with what may become an invaluable climate intervention option. I hope readers might also find something beyond that narrative, as I think the characters within can be guides to a richer worldview. "There is a particular appreciation of the earth-system that can be gained only by imagining how it could be changed," wrote Morton. Herein, then, lies a path from bewilderment at the earth's complexities, and the myriad ways humans interfere, to wonderment at the earth's systems, and how humans can better replicate them—a path that might lead to opening doors and looking at the future differently.

FALSE COMPETITION

I N 2007, RICHARD BRANSON held a press conference to announce a competition that would see the winner earn $25 million in prize money and, potentially, the satisfaction of saving humankind itself. It was called the Virgin Earth Challenge: an invitation to scientists, engineers and other innovators to devise a way to remove at least one billion tons of carbon from the earth's atmosphere per year in a commercially viable way. After Branson spoke of the vital importance of the award's intent, he challenged those who might be capable of such a feat to put their minds to it. "Most of us have only really encountered the concept of a planet under threat in science fiction films. The plot is often that no one believes the threat until it's almost too late and then a superhero steps in to save the day," said Branson. "Today we have the threat—we still have to convince many people that it is indeed both urgent and real—and we have no superhero. We have only our ingenuity to fall back on."

Seated next to Branson at the announcement was former U.S. vice president Al Gore. Removing carbon dioxide from the atmosphere was not something Gore had given much thought to prior to teaming up with Branson. In his film released the previous year, *An Inconvenient Truth*, he argued that the only missing ingredients for tackling climate change were the personal and political will to eliminate emissions. When the film tour made a stop in London, Branson was invited to attend the U.K. premiere but couldn't make it. So instead, Gore approached him to ask for help in stemming the climate crisis. Branson, who is known for being accessible, was reached in the bath and informed that Gore would like to make a house call to share his perspective directly. "He did his whole inconvenient truth thing in my living room," Branson recalled. The visit fundamentally changed the worldview of the founder of the Virgin multinational conglomerate, which had over 200 branded companies at the time. "Not only was it one of the best presentations I have ever seen in my life, but it was profoundly disturbing." He had previously been a climate change skeptic, but doubts had begun to surface after he attended a meeting with scientists that was convened on his behalf in response to a proposal he had made to build a refinery (so he could lower fuel prices for his airline business). Now, after seeing Gore's presentation, he was left with a firm sense of urgency that immediate personal action was required; and characteristically for someone with little tendency to deliberate, he sprang into action, seeking solutions to the intractable problem.

He began his quest by seeking out a better understanding of how Earth actually functions, hoping that might provide some clues to how best to intervene. A key influencer was a British scientist named James Lovelock, who had helped others reflect more deeply on humans' relationship with their planet through his Gaia Hypothesis, an explanatory theory for how Earth functions as a self-regulating system. Lovelock was working for NASA as a consultant in the 1970s when he was asked how one might determine if another planet, Mars for example, harbored any life. Rather than inventing instruments that could directly probe for microbial life on the red planet, as others did, he drew on experience he had designing gas detection devices, the most well-known of which could detect molecules in the atmosphere at the previously unfathomable scale of parts per billion. That led Lovelock to come up with a novel approach for detecting life on other planets: comparing atmospheres. Earth's atmosphere wasn't magically created by random geological events that happened to make the planet habitable for humans from the outset over four billion years ago; it became habitable over time as life-supporting gases like oxygen became prevalent in the atmosphere. In a profound observation, Lovelock surmised that everything that has ever breathed, grown and decayed on Earth has played a role in constructing its atmosphere, interacting with inorganic surroundings in a self-regulating system. Components of that system both influence and respond to others, including the global temperature. (Meanwhile, on Mars, recent observations had shown the atmosphere to be indicative of a "dead planet," made up almost entirely of carbon dioxide, a composition toxic to life.)

When Branson sought out Lovelock for guidance, he was left with not only a better understanding of how the earth's natural world is composed of feedback systems that create stability but also more reinforcement that those systems were being overwhelmed by human activity. Increasing concentrations of carbon dioxide and other greenhouse gases in the atmosphere were having the effect of wrapping the earth in an increasingly thicker blanket, with destabilizing climate implications. Branson took his newfound concerns to a meeting with Virgin's corporate and brand development director at the time. The two ruminated on what the company could do differently and decided to overhaul the way it operated. They believed the company could create wealth and jobs and safeguard the planet at the same time—it would be called Gaia Capitalism.

Another revelation followed: "I was lying in the bath and I thought, 'Screw it. We make a lot of money out of the airline businesses and the train businesses. Let's just tie all that money, for the next ten years, into trying to develop fuels that don't damage the environment.'" By chance, Bill Clinton phoned shortly after and asked if Branson had considered making a pledge at the upcoming 2006 Clinton Global Initiative annual meeting. To Clinton's delight, Branson informed him of his revelation. On day two of the gathering, after $2 billion in commitments had already been announced, aimed at "creating a small piece of common ground in a contentious world," the host took the stage to introduce the centerpiece of the event. When Branson made his way to the stage, he pledged that all profits from the Virgin Group's transportation businesses would be invested in the development of cleaner fuels over the next ten years. "Our generation has inherited an incredibly beautiful world from our parents and they from their parents," he told the audience. "We must not be the generation responsible for irreversibly damaging the environment."

Swarmed by press following him out the door after his announcement, Branson suggested that profits dedicated to developing new fuels could total over $3 billion. Perhaps the number was inflated in the excitement of the moment; the actual amount, spent mostly on developing biofuels, has been about 10 percent of that estimate (Branson has blamed the shortfall on the 2008 financial crisis). Another explanation might be that efforts were largely aimed at the airline industry, one of the most difficult industries to decarbonize—jet fuel is one of the most concentrated forms of energy available. That was illustrated two years after Branson's announcement, when a Virgin 747 flew from London to Amsterdam fueled partly by oil from coconuts. In trademark fashion, Branson drew publicity by joining the flight's captain for a photo of them drinking coconuts through a straw. Coconuts made up less than 5 percent of the fuel, and that still required 150,000 coconuts. For perspective, replacing the entire fuel for the one-hour flight would require about three million coconuts, hardly practical.

Branson was looking for bigger ideas though. Following the Clinton Global Initiative announcement, he was at Necker Island, his now wind- and solar-powered retreat in the British Virgin Islands. A climate change discussion was taking place on a TV in the room, and Branson expressed disdain with the situation to his wife, Joan, who he describes as a matter-of-fact Glaswegian who keeps him grounded.

"There must be some genius out there who can remove the carbon from the atmosphere," said Joan.

"There probably is, but how do we find him?"

"Why don't you offer a prize?"

Branson turned and stared at her as an epiphany dawned on him. He had read Dava Sobel's book *Longitude*, about the prize offered by the British government in 1714 to allow seafarers to navigate with the help of geographic coordinates. The prize was ultimately awarded to a clockmaker named John Harrison, who came up with an unexpected solution by designing a clock that could keep the time at a reference location, which could then be compared to local time determined by the sun's position. The timekeeping solution came as a surprise to the scientific establishment, which had expected an astronomical solution— clocks were deemed to be too unreliable and cumbersome—and the British Board of Longitude initially refused to recognize it (Harrison, who was twenty-one when he first started working on a solution, received his final payment when he was eighty). The eventual vindication of Harrison's method made a strong case for the use of inducement prizes as an alternative to more conventional approaches, like government grants, that might be too narrow in scope to contemplate outside-the-box solutions.

Another prize had a more direct impact on Branson. The Ansari X Prize—itself inspired by the Orteig Prize, which chal- lenged aviators to fly from New York to Paris—was a competition offering $10 million to the first privately funded launch of a crewed, reusable spacecraft that reached suborbital altitude

100 km above the earth's surface twice within two weeks. It was awarded in 2004 to a spacecraft funded by Microsoft co-founder Paul Allen. That same year, Branson licensed the winning technology and started up Virgin Galactic with the aim of taking tourists to space by 2009. (Virgin Galactic was always about a year away from doing so until the moment finally arrived in 2021.) One takeaway from the Ansari XPrize was the opportunity to leverage funds for much greater potential. For example, donations made to fund the Ansari XPrize were used to pay the premium on a "hole-in-one" type insurance policy that would guarantee a $10 million payout if the challenge was met in time. In turn, the competition spurred over $100 million in investments in competing technologies, according to organizers. Spin-off competitions later attempted to spur innovations ranging from reversing ageing in mice to designing a tether to space that would act as a space elevator.

In Branson's mind, there weren't any challenges more befitting of an inducement prize than carbon removal. Moving quickly per his nature, he began recruiting help to establish the competition and evaluate the viability of entries. After his initial meeting with Gore, who later agreed to be a judge, Branson began reading books on climate change. One, *The Weather Makers*, written by an Australian scientist named Tim Flannery, had a profound effect on him. The book's author received a call one day from Branson's office inviting him to Necker Island to address the leaders of Virgin companies on climate change. Subsequently, Flannery was invited back to the island to talk about the Virgin Earth Challenge: "I talked very extensively with Sir Richard Branson about drawdown [of carbon dioxide levels]. He had a very pessimistic view of the outcome of the [United Nations] climate meetings and felt we needed an insurance policy." Flannery agreed to join as a judge.

Other climate luminaries were persuaded to join as judges, including NASA climatologist James Hansen, whose 1981 climate change predictions led to a seminal U.S. congressional testimony later that decade; and former U.K. diplomat and environmentalist Sir Crispin Tickell, whose 1977 book about climate change led to influencing Margaret Thatcher on the subject. The most controversial selection was James Lovelock. Gore had voiced misgivings about Lovelock's pessimistic view of the ability to bring nature back into balance. Indeed, Lovelock characterized Branson's concept as futile: "Trying to restore the earth by removing one or even two billion tons a year is a bit like trying to bail out a leaky rowing boat with a teaspoon." And hubristic: "I would sooner expect a goat to succeed as a gardener than expect humans to become stewards of the Earth." Nevertheless, he found the challenge intriguing and agreed to participate.

The prize would be awarded to any individual or group developing a technology capable of, or appearing to be capable of, removing carbon dioxide or other greenhouse gases totaling at least one billion metric tons per year of carbon equivalent (which is equivalent to 3.7 billion metric tons of carbon dioxide after factoring in the weight of the two oxygen atoms for every carbon atom) over the course of at least ten years. The contest would be open for five years and the prize would be awarded by the panel of judges. Branson and Gore posed for pictures at the announcement, Branson gleefully tossing a globe in the air like it was a plaything while a much more circumspect Gore looked on. "It's a challenge to the moral imagination of humankind to actually accept the reality of the situation we are now facing," said Gore.

Within a month of the announcement, over 5,000 applications were received. Overwhelmed by the response, the organizers asked a team at the University of Cambridge to help comb through the entries to evaluate whether any held breakthrough potential. Four years later, after receiving over 5,000 more applications, interest had been whittled down to about 2,500 formal submissions. They contained a vast array of different ideas for removing carbon. "They have fundamentally altered my perception about how we might respond to the climate crisis," wrote Tim Flannery in his 2015 climate change book, *Atmosphere of Hope*. In fact, he attributed the book itself to his involvement in the competition, the process having given him a source of inspiration after being sidelined by despair while witnessing the problem worsen following the release of *The Weather Makers* a decade previously.

The entries also revealed how far away a viable solution remained. Not only were many of the entries nonconforming but, of greater concern, none were remotely close to being ready for large-scale deployment. Skepticism grew within the group administering the competition that the breakthrough solution envisioned by Branson could be found. "Awarding one of these projects $25 million now won't do the planet much good when what we've actually learned these past years is that it still might take as much as $2.5 billion to develop a machine or process to suck the carbon out of the air, and keep it out," said a spokesperson in 2010.

Regardless, eleven finalists were announced in 2011, mostly nature-based approaches to removing carbon dioxide. Three proposed using biochar, a charcoal-like substance that can be created by burning decomposing biomass, such as agricultural and wood waste, in an oxygen-limited process that produces almost no polluting gases. The heat created during the making of biochar can be used as an energy source, and the biochar itself can be mixed into soil to improve many of its functions, including greater carbon dioxide uptake. Another finalist proposed enhancing the carbon uptake of subsurface minerals, a process called enhanced weathering, by bringing minerals like olivine to the surface. The olivine would then be mined, crushed and spread over beaches and other areas that would agitate it further, thus accelerating the process of removing carbon dioxide from the atmosphere. Another finalist proposed capturing the carbon dioxide emitted from combusting biomass to produce electricity or other products of value and storing the carbon dioxide underground, which is referred to as bioenergy with carbon capture and storage, or BECCS. (One finalist proposed changing the management practices for grasslands to sequester more carbon dioxide in the soil but was later disqualified due to concerns with the permanence of the carbon dioxide sequestration.)

The approach taken by the remaining five finalists entailed the direct removal of carbon dioxide from the atmosphere using chemical technologies and industrial equipment, generally referred to as direct air capture (DAC). The concept is nothing new; it was first deployed in the early 1600s on the Thames River in London. The inventor, Cornelis Drebbel, had been given room and board and a regular stipend by King James I to establish a laboratory in a royal palace and design apparatuses that replicated phenomena found in the natural world—a perpetual motion machine, a self-regulating oven, a magic lantern, a thunder and lightning device, an air-conditioning unit—that served to entertain the royal court. Drebbel thought he might find broader demand for a contraption that could travel underwater to sneak up on unsuspecting ships during naval warfare. He even took his royal patron for a tour in one oar-powered, leather-hulled submersible that could travel from Westminster to Greenwich, but he was ultimately unsuccessful in finding a willing backer for his invention. In the process, however, he demonstrated that through a straightforward chemical reaction—it is believed that he heated potassium nitrate in a metal container, producing oxygen and potassium oxide that bonded with carbon dioxide— enclosed air, like that in his submersible invention, could be scrubbed of carbon dioxide. Four centuries later, and in need of scrubbing a vastly larger volume of carbon dioxide, we may come to rely on the same concept.

In 2018, the Intergovernmental Panel on Climate Change (IPCC) published a 630-page special report, "Global Warming of 1.5°C." Much of the report described the many disconcertingly compelling reasons to limit warming to that level. It also found that all the pathways considered in attempting to limit warming to 1.5 degrees Celsius include the use of carbon removal. And not a trivial amount: somewhere between one hundred billion to one trillion metric tons over the course of the 21st century. To meet that target would require deployment of technologies developed by entrants in the Virgin Earth Challenge on a massive scale.

Despite the report's assurance that significant demand should be expected for carbon removal technologies, the Virgin Earth Challenge was terminated the year after the report was released, in 2019, and no prize money has ever been awarded. Those working on potential solutions were informed that "the market conditions necessary to support commercial and sustainable investment in the relevant carbon removal techniques were not foreseeable." One of the terms of the competition was that a successful entry be commercially viable based on future markets. But the purpose of using incentive prizes is to accelerate research and development in the absence of existing market demand or other support, which, if successful, might then spur the creation of future markets. For the Virgin Earth Challenge, however, it was insufficient for entrants to envision a future carbon removal market, even one shown to be imperative by the IPCC report.

Yet some entrants remain undeterred. They continue improving on the technologies they've developed and, with the help of others, are building markets. This book is the story of four of them and the technologies they are developing to scrub the sky of excess carbon dioxide.

THE ORIGINAL
SKY SCRUBBER

O N A WARM SUMMER EVENING in 1992, Klaus Lackner and a friend were reminiscing over a beer on the back deck of Klaus's home in Los Alamos, New Mexico, about sharing a rare and sought-after computer in a campus research lab in the early 1980s. As the evening wore on, the pair turned their conversation to some of the big ideas people contemplated more frequently in the past, when the potential for unintended impacts was less of a concern. A few beers later, the conversation evolved into their own big idea: What if some of the world's biggest problems could be solved with large colonies of self-replicating machines?

Looking out over the barren, mesa landscape unfolding before them, they envisioned the machines digging up raw dirt, transporting it on ceramic tracks, producing metals from electrified ovens, manufacturing machine parts and solar panels, and building more of themselves, all around the clock, in a self-contained system. This particular location that they gazed upon, near former atomic bomb testing grounds, would potentially be large enough to power all of the United States, while other colonies could provide additional services, like cleaning up the excess carbon dioxide building up in the atmosphere. An extremely large processing system would be required to remove carbon dioxide at a sufficient scale to stop global warming, a task that could be carried out by a colony dedicated to transforming the planet-warming gas into mountains of rock.

When the two friends met up for breakfast the next morning, they decided that the previous night's discussion was not actually all that crazy. They returned to their regular jobs and, in their spare time, began testing the theory for obvious bottlenecks. The most likely in their estimation was the reliance on raw dirt as the main input for the manufacturing to be carried out by the machines. The pair concluded that despite many of the essential elements, like aluminum and iron, being less readily available than in conventional source materials, the machines would just have to work harder to separate the dirt into its elemental constituents. Meanwhile, manufacturing inputs requiring rarer elements, such as copper and platinum, would have to be redesigned or eliminated, and other elements unavailable in dirt—hydrogen, carbon, nitrogen—would be captured from air and rainwater. Any excess materials could then be used for non-critical items.

To capture excess carbon dioxide from the air, the machines would be tasked with constructing carbon dioxide extraction units. The required inputs would be minimal: a metal oxide, possibly magnesium or calcium, and air. Assuming enough wind was present to move air through the extraction units, the metal oxide would bond with the carbon dioxide in the air, producing carbonates similar to naturally occurring minerals. Klaus theorized the total output volume of minerals created by the carbon capture would create a layer roughly a meter and a half thick if it were to cover an area the size of New Mexico.

Klaus called the machines auxons—their name derived from the Greek word for "to grow"—and they would be small enough that their own replication wouldn't hinder the growth rate of the system. Specialized tasks would be assigned—digger aux-ons, transport auxons, production auxons, assembly auxons, administrative auxons—that could be readily automated. Their relatively simple and efficient design risked becoming a bottle-neck, though, as a growing, self-contained system would still have to be intelligent enough to repair itself as human interven-tion ceded control to the self-directed, simpleminded automatons. Recovery procedures could be implemented, but what would happen when reboots would not suffice? The pair concluded that auxons would at least be given enough sense to know when to retire and make their way to the recycling bin. And to prevent any wayward auxons from stepping outside their assigned task, humans would still exert ultimate control with the ability to turn the power supply off.

Thus satisfied that no obvious roadblocks stood in the auxons' way—the concept was even tested with imaginary disasters—Klaus and his friend Christopher Wendt co-authored a paper in 1994 laying out their scheme. The next step: spending more time researching the alchemy of dirt. But Klaus eventually moved on to other things, and no auxons have ever roamed the New Mexican mesas. "That was a wild and woolly paper," says Klaus now. But in the process of thinking through his quixotic vision, he learned a lot. "The question is: Can it be done? We concluded there are some interesting aspects to this problem and the abilities [the auxons] would give us, but it was put aside because I didn't think this was the next big thing—it was the next, next, next big thing. Mean-while, what do we do about that carbon problem?"

Klaus grew up in the small, picturesque German city of Hei-delberg, one of the country's scientific hubs. As a teenager, he read a book called *One Two Three... Infinity*, written by the-oretical physicist George Gamow. It laid out math and science concepts, "explained for the layman," and has been credited with setting many influential scientists on their career paths. Klaus was similarly influenced to study theoretical physics at his hometown Heidelberg University, one of Germany's oldest and most reputable. He felt stymied by the German seniority system that prevailed at the time, though, so he sought out greater possibilities in the U.S., completing a postdoctoral fellowship at Caltech, a research university near Los Angeles. "I loved coming to Caltech. Even if you were a young postdoc, you were taken seriously. It was strictly just the message that you had to defend and explain. I found that exhilarating." When he was subsequently offered a faculty position in Germany, he felt that the meritocracy governing relationships in his new home afforded him greater opportunity and so he remained. He accepted a position at the Los Alamos National Laboratory in 1983, which had been established during World War II to design nuclear weapons as part of the Manhattan Project, and stayed there for two decades, working on supernovas, bombs, neural nets and laser fusion. Following the end of the Cold War in the early 1990s, the lab shifted focus to researching solutions to other security threats, including climate change.

The auxon paper had planted the seed for Klaus of capturing carbon dioxide from the atmosphere, but at the time he believed the bigger challenge would be finding a place to safely store all the carbon dioxide he foresaw being emitted in a world that continued burning fossil fuels. "Carbon dioxide piles up in the atmosphere and it is cumulative—it is a stock problem, not just a flow problem—and sometime in the 21st century we would be running into a brick wall," he said about his reasoning at the time. Klaus estimated then that annual carbon emissions could average forty billion tons per year over the next hundred years, which would equate to four trillion tons of carbon to be mineralized in the ground (roughly fifteen trillion tons of carbon dioxide). Any capture technology, whether it be from point sources like smokestacks or pulling it directly from the air, would get better over time; but for storage, the inverse was true. "Storage gets more and more scary the more you store," said Klaus. "If you ask me,

'Can I store a ton?' Of course. 'A million tons?' Not a problem. 'A billion tons?' Take a deep breath and we can do that too. But we're talking trillions of tons."

That left Klaus convinced that renewable energy would have to replace fossil fuels, but he was concerned that it would take too long to complete the transition and the dilemma of what to do with excess carbon dioxide in the atmosphere would remain. Nobody was researching how best to extract carbon dioxide from the air to limit its buildup, because it was viewed as an impractical undertaking given how diluted carbon dioxide is in the air: roughly 365 parts per million at the time, or 0.04 percent. But Klaus looked at the problem differently. His moment of revelation came when someone asked him to look at a proposal to build wind towers that, instead of simply converting the kinetic energy of wind to electrical energy, would pump water high up into the air, where it would evaporate and cool the air. The cold air would then fall back to earth, where the turbines would collect the kinetic energy of the falling air. Klaus was unconvinced that it was a practical idea, but it got him thinking about the amount of energy represented by a cubic meter of air.

He first calculated how much kinetic energy a wind turbine sees when a cubic meter of air moved through it. He then calculated how much energy is released when burning an amount of fossil fuels that produces as much carbon dioxide as is present in that same cubic meter of air. It turned out the energy extracted by the wind turbine was 500 times *less* than the energy expended from the burning of fossil fuels that put the carbon dioxide into the air. So Klaus asked, "If that's true, and we know we can pull kinetic energy out of the air, why couldn't we pull CO_2 [carbon dioxide] out of the air?" Expressed differently, if a wind turbine were to sweep an area equivalent in size to the floorspace of a Boeing 737, it would produce a certain amount of energy. For an air capture unit to remove the carbon dioxide produced by burning fossil fuels that generated the same amount of energy, the area holding that amount of carbon dioxide would be more like the size of a couple of windows on the airplane. Yet, despite the fact the wind turbine must sweep a much greater area to generate the equivalent amount of energy, it was economic to do so. Therefore, it followed that it would also be economic to burn fossil fuels *and* capture the associated carbon dioxide. "That was the starting point," said Klaus.

He knew from high school chemistry that a strong alkaline solution like sodium hydroxide or calcium hydroxide would bind with the carbon dioxide in the air. And the cost of a wind turbine could be used as a rough proxy for building a structure designed to convert air into something. Putting all those inputs together, his model spat out a cost estimate of fifty cents per ton of carbon dioxide removed, but he was using a wind turbine as a stand-in for the technology that would have to be built, so he factored up his estimate by ten to be safe, to five dollars per ton. The larger cost not yet factored into his calculations would be in the backend: prying the carbon dioxide away from the chemical sorbent used to pull it out of the air. After doing some more math (in Klaus's jargon: "the binding energy is only logarithmically dependent on the concentration"), he concluded there wasn't all that much more energy needed to make it work.

A few months later, he was at home working in his office when his daughter Claire appeared, asking for help coming up with an idea for her eighth-grade science fair entry. Presumably many scientists are susceptible to becoming overly involved in their kids' science projects. For Los Alamos scientists, it was instantly competitive. "Why don't you pull CO_2 out of the atmosphere?" he suggested. "You could try using sodium hydroxide." Claire went to the local pet store and purchased an aquarium pump that injects air into fish tanks to keep them oxygenated. She then filled a test tube with a sodium hydroxide solution and attached the aquarium pump to circulate air through the tube. With carbon dioxide being an acid, like vinegar, and sodium hydroxide being a base, like baking soda but much stronger, the two bind together in an acid-base reaction when put in contact. The result is that carbon dioxide binds with the sodium hydroxide and leaves the air; over half of it in the case of Claire's experiment. The entry won her first prize at the 1997 science fair. "I was surprised she pulled it off as well as she did," said Klaus. "It made me feel that it could be easier than I thought."

A couple of years later, Klaus was being interviewed about removing excess carbon dioxide from the atmosphere. "This must be a horribly difficult problem," stated the interviewer. "Actually, no it's not," replied Klaus. "Even my daughter can do it—she did it for a science fair." He decided to publish a paper, along with two Los Alamos colleagues, that explored whether mitigating greenhouse gas buildup by extracting carbon dioxide from the air was worth further study. Several carbon removal options were explored, like using certain minerals to enhance the natural mineralization of carbon dioxide, a task he had envisioned auxons taking on, but he focused on direct air capture. "It is known from simple experiments that calcium hydroxide solutions are very efficient in collecting carbon dioxide from air," stated the 1999 paper. "Simply bubbling air through a wash bottle removes the bulk of the carbon dioxide." The citation for the observation was "Claire N. Lackner, private communications."

The paper concluded that despite the inherent disadvantage of not capturing carbon dioxide directly from more concentrated emissions sources like a coal-fired plant emitting flue gas, there were also advantages to direct air capture. One was the flexibility in scaling to a large size to take advantage of economies of scale, and another was the greater siting options to take advantage of existing infrastructure and lower transportation costs for sequestering the carbon dioxide belowground. The biggest challenge would likely be the amount of energy required. "We needed to come up with a shape where you don't have to have an aquarium fish pump driving all the air through the system but to have the wind just deliver the air and pass it through the collector," said Klaus.

Other scientists were not convinced that it was possible. When one of those scientists, a senior research engineer at the Massachusetts Institute of Technology (MIT) named Howard Herzog, first came across Klaus's paper stating that the cost of pulling carbon dioxide from air could eventually be less than 25 cents per gallon of gas ($25 per ton of carbon dioxide), he was incredulous. Herzog produced a white paper for the U.S. Department of Energy that concluded that direct air capture couldn't work. The critique gave Klaus the sinking feeling that he might have made a mistake. He read the paper again. "If Herzog was right, Claire's experiment couldn't possibly have worked, so I sat down to find the mistake." It took another half a day to find it. Underlying the work supporting Herzog's paper was the use of an alkaline solution to bind with the carbon dioxide that was equivalent to using seawater, which would indeed make the economics dismal compared to using a much stronger alkaline solution. "The reason I was so sure that he had to be wrong was that Claire had done her experiment, and I had seen it, and there was nothing in that experiment that would have worked if he had been right," said Klaus. But debating the laws of thermodynamics was not going to convince skeptics. For that, he would have to build an actual direct air capture unit.

BIOSPHERE 2 IS A THREE-ACRE greenhouse situated outside of Tucson, Arizona, that was built in 1991 to house an experiment that would test whether life could be sustained while sealed off from the earth's atmosphere. It was funded by an oil and gas billionaire named Ed Bass, who was interested in testing the possibility of space colonization. Inside were various biomes—rainforest, wetland, savanna, desert, ocean—and a farm, all designed to be microcosms of ecosystems found in Biosphere 1, Earth. Eight months after locking eight individuals in the enclosure for a two-year trial, however, problems started to develop: the Biospherians started losing weight, and oxygen levels of the Biosphere 2 atmosphere dropped from 21 percent, equivalent to the outside air, to 14 percent. A world-renowned scientist named Wally Broecker, who had co-authored a book entitled *How to Build a Habitable Planet*, was called in to problem-solve (in 1975, Broecker also authored the first paper to use the term *global warming*).

Broecker discovered several reasons why oxygen was being suppressed, including an overabundance of bacteria-rich soil spread throughout the enclosure that was consuming a surprisingly large volume of oxygen and, correspondingly, respiring a lot of carbon dioxide. The most enlightening outcome for Broecker, though, was the discovery by one of his graduate students that carbon dioxide levels were being suppressed from being even higher. The student noticed rinds on the unsealed concrete used in the foundational structures. A reaction was occurring between the carbon dioxide and the calcium hydroxide contained within the concrete to form calcium carbonate, a powdery limestone. Some calculations confirmed that unsealed concrete was indeed accounting for the missing carbon dioxide. The ability to track carbon inside the closed system had provided a valuable demonstration of the accelerated mineral sequestration that Klaus was theorizing about around the same time but that was difficult to observe outside in the real world.

To capture the excess carbon dioxide that wasn't being mineralized by the unsealed concrete, a scrubber was installed that used sodium hydroxide to react with the carbon dioxide, effectively a larger version of Claire's experiment. It wasn't installed at a sufficient scale, though, so eventually oxygen was trucked in to allow the Biospherians, who overcame many other challenges, to complete the two-year trial. After improvements were made based on extensive research, a second mission began in 1994 with an intended duration of ten months, but bigger challenges loomed. The billionaire behind the project, Ed Bass, had become concerned with ballooning expenditures and brought in an investment banker named Steve Bannon—yes, the same Steve Bannon who later became Donald's Trump's adviser—to take over and help alleviate the financial woes. Bannon tried to find a way to earn some revenue from Biosphere 2—he even looked at developing a Biosphere 3 in Las Vegas that would include a casino and resort—but eventually turned to reducing costs. The management team overseeing the biosphere was fired shortly thereafter. A member of the original mission named Abigail Alling, who was acting as a safety consultant for the second mission, witnessed Bannon's takeover. The investment bankers showed up outside the sealed doors in limousines, temporary restraining orders in hand, to take control of the project with police support. The office locks were changed, and those who had been overseeing the mission from the outside were prevented from receiving any further information about what was occurring within the sealed enclosure or communicating with the Biospherians inside.

Alling, who has since gone on to head up the nonprofit Biosphere Foundation, was concerned for those inside. She believed they deserved to know about the abrupt changes and that their safety was compromised by the new management team's lack of familiarity with the facility's complexities. She raised her concerns in the form of a five-page statement that she presented to Bannon. He responded by threatening to "ram it down her fucking throat." (He initially denied having made the threat, along with other vitriol directed at Alling, but it was later revealed in court that he had been tape-recorded by the director of engineering, who had hidden a recording device in his underwear.) With all communication systems hijacked, only one means of communicating with the crew inside remained: entering Biosphere 2. Abigail snuck onto the grounds with a team member from the first mission in the middle of the night and opened the doors, breaking the seal. The crew inside decided to stay, but their confidence in their outside handlers was severely hampered. The mission was terminated prematurely several months later.

In June 1994, the entity that owned Biosphere 2 was dissolved and management of the facility was handed over to Broecker's employer, Columbia University, to use it for climate change research. Before his exit, Bannon gave an interview with C-SPAN—which now seems like a deepfake given his more recent opposition to climate action—in which he said that Biosphere 2 was more useful as a laboratory for expanding knowledge of the impacts of climate change than for gaining any knowledge that might be useful for extraterrestrial living: "The power of this place is allowing those scientists who are really involved in the study of global change, and which, in the outside world—or Biosphere 1—really have to work with just computer simulation, this actually allows them to study and monitor the impact of enhanced CO_2 and other greenhouse gases on humans, plants, and animals."[1]

Broecker and his Columbia colleagues similarly saw an opportunity to use Biosphere 2 to study the impact of rising atmospheric carbon dioxide levels and continued running experiments with Biospherians no longer present. The Los Alamos National Laboratory was also looking for new areas of interest, and Klaus finagled an invitation to attend a conference at Biosphere 2 put on by Columbia in 1999. Broecker had first come across Klaus at a climate conference in Ottawa the year prior, where Klaus gave a presentation summarizing his thoughts on accelerated mineralization of carbon dioxide—mining billions of tons of crushed rock that could absorb carbon dioxide from the atmosphere, producing mountains of carbonate in the process.

---

1 After the 2016 U.S. presidential election, Broecker became alarmed when Bannon, now in a very influential position, opposed efforts to address climate change, reportedly influencing President Trump to pull the U.S. out of the Paris Agreement on Climate Change. Broecker tried contacting Bannon to ensure he understood the scientific consensus but never heard back.

Nothing was said about auxons, but Broecker still thought he was nuts, just "another guy from our national laboratories looking for a way to justify his continuation." Klaus stood his ground: "I'm not easily intimidated when it comes to these things. We had a good, aggressive discussion."

Despite the pair's seemingly divergent views, or perhaps because of them, Klaus made a good impression at Biosphere 2 and was invited to join the scientific advisory committee looking to create a world-class research facility that Broecker also served on. As the two scientists subsequently got to know each other in the more collaborative setting inside the glass walls, Broecker began to see Klaus's grand ideas in a more tantalizing light. "Here was one very bright man," said Broecker.

He began trying to lure Klaus to join him at Columbia University. The enticement of university life eventually overrode the Lackner family's attachment to living in Los Alamos, and Klaus agreed to interview for the position. Academic interviews often feature the candidate talking about a topic of interest, and Klaus gave an in-depth presentation on capturing carbon dioxide from the air. There was nothing revelatory about the feasibility of doing so; scrubbers had been used in submarines, space shuttles and, of course, in Biosphere 2 itself. What captured his listeners' attention most was his explanation for how it was economically feasible to scrub the atmosphere of carbon dioxide despite the seeming impossibility posed by the diluteness of carbon dioxide in air. "He has the theoretical physicist's penchant for attacking problems from a base of first principles, but he is patient with people who are less well-grounded in those principles—thermodynamics, say—than he is. The lack of arrogance in somebody so obviously smart is charming," Broecker later wrote in a book, *Fixing Climate*, jointly written with Robert Kunzig. Another professor at Columbia, Julio Friedmann, now one of the world's most authoritative experts on carbon removal, has also shared his effusive praise, telling the *New Yorker*: "Klaus is, in fact, a genius."

Broecker was keen to see Klaus's concept prototyped and also knew that Klaus would need some help pulling it off, including someone with a more hands-on approach to complement him. He had just the person in mind, someone Klaus had worked closely with at Biosphere 2 who, conveniently, was already intimately familiar with Klaus's views on capturing carbon dioxide.

That person was Allen Wright, an engineer who took an unconventional path to becoming one. Before immersing himself under the glass pyramids and geodesic domes in 1999, Allen had spent decades taking machines apart and putting them back together again. Allen grew up in Detroit, where his dad worked for Ford as an automotive engineer and his mom taught diesel engine mechanics to servicemen during World War II. Outside their house, cars could always be found sitting on blocks. Allen and his two brothers were frequently left on their own, and so they built things using the parts strewn about, like a propeller-powered sled out of a lawnmower engine. None of them studied engineering at university, yet they all became engineers. Allen also loved exploring nature, whether traipsing through wetlands and marshes or seeing through the eyes of Jacques Cousteau on TV. His eighth-grade biology teacher fostered his fascination with ecology, taking him on excursions to explore pondlife or to bring alive ancient life by looking for trilobites in old quarries. After completing school, Allen bounced around between university and construction jobs before he eventually completed a degree in fisheries biology. But the only job available upon graduating entailed sitting on a rock in the Aleutian Islands counting seals, so he returned to construction.

On a whim, Allen and his new wife moved to the Cayman Islands, where she found work as a social worker. As Allen deliberated about what he would do, a small submarine parked at the nearby pier caught his eye. He tracked down the owner and asked if he could be of service in any way. The interview was brief. "Do you smoke?" (Conveniently Allen had recently quit.) "Do you have a college degree?" (It had taken Allen twelve years, but he did have one.) He was offered his dream job: submarine pilot. From there, Allen moved on to a startup in Vancouver that was manufacturing small, transparent submarines to be used for research and tourism. That led to a position with the University of Hawaii, building out their submersible research program. After a decade there, it was time to return to the mainland. At his going-away party, someone suggested he speak with a professor at the university who was responsible for the ocean biome at Biosphere 2 and was working with Wally Broecker to better understand what was happening with carbon dioxide levels within the enclosure. As outside scientists working with staff employed directly to run Biosphere 2, they were having trouble completing the research they were most interested in. They wanted their own guy on-site, someone with a track record of converting complex ideas into simple equipment.

When Allen arrived at Biosphere 2, it was only partially sealed off from the outside world. A ventilation system allowed some exchange of air with the outside biosphere to avoid large build-ups of carbon dioxide at night, when the vegetation respired, and drop-offs during the day when it photosynthesized. Each biome could be controlled separately, so scientists could test, say, how changing carbon dioxide concentration might impact vegetation growth or coral growth in the ocean. Every month Allen and Klaus would talk about the theory behind the operation and the technical elements needed to make it work. "He would talk to me about Rho v squared and delta Ts and R naughts," said Allen. "He speaks an odd language, Klaus does. He thinks in mathematics— it's how he formulates ideas. They're all mathematical constructs that he translates into language. He's a remarkable person." Allen would try as hard as he could to keep up with what Klaus was saying. And then the tall German would speed off at twenty miles per hour to talk to someone else and Allen would put his head on his desk to take a nap.

While Allen and Klaus's time together at Biosphere 2 would come to an end in 2003, when Columbia ceased managing the facility because of budgetary concerns, the pair had complemented each other well and would soon join forces to seek out a prototype for taking on a much larger task.

GARY COMER WAS AN AVID sailboat racer who started selling sailing equipment out of a basement apartment in Chicago in the 1960s with some friends. They grew it into a large mail-order apparel company called Lands' End (a printer error placed the apostrophe in the wrong spot, but Comer decided to leave it there) before selling it to Sears in 2002 for $2 billion. A year prior, Comer had sailed the Northwest Passage on a whim. He had been inspired to do so by early explorers he'd read about. However, his trip went so smoothly in ice-free waters that he had difficulty reconciling his experience with crews like the 1845 Franklin expedition, where all 129 men perished after being trapped in ice for a year and a half. Comer began researching climate change and became convinced that additional scientific work was required to avert a crisis. He asked around to find someone he should speak to and was told to get in touch with Wally Broecker.

Comer mailed a letter to Broecker introducing himself and describing his disconcerting trip through the Northwest Passage. He asked if Broecker would join him in Chicago for a meeting as soon as possible. Broecker wrote back to say that his teaching responsibilities prevented him from doing so for at least a couple of weeks. Comer, who had recently been diagnosed with advanced prostate cancer, wasn't prepared to wait. He phoned Broecker to tell him he would be there shortly to meet for breakfast and explore what could be done to combat climate change. Having both grown up in Chicago around the same time, the two formed an instant connection. Broecker quickly introduced his new friend to his colleague: "Klaus is the most brilliant person I have ever dealt with. I always say the world is really lucky that

Klaus is putting his energy into air capture." When the three met, they talked about advancing Klaus's direct air capture concept, and the pair from Columbia resolved to provide a formal pitch to Comer in short order.

Broecker and Klaus agreed that Allen Wright was the best candidate to build a prototype. Allen was agreeable when asked, and the trio, along with Allen's brother Burt—a former firefighter who enjoyed mechanical tasks and now ran an engineering firm that designed sprinkler systems—gathered in a conference room at a New Jersey airport to pitch Comer on funding the creation of a direct air capture unit. Allen had some familiarity with the mechanical components that might be required for an air capture system, but his head was spinning with Klaus's ideas, so he assumed he was going to be a fly on the wall at the meeting. He'd be listening to the finest academic minds he had ever encountered convince the wealthiest person he might ever meet to help start up a new industry that might play a role in solving the greatest challenge to ever face humankind. But before Comer arrived, Broecker turned to Allen: "By the way, you'll be doing all the talking."

"So, why are we here?" asked Comer when he sat down across from the Wright brothers.

I don't quite know, thought Allen, taking a moment. "Well, here's what I think I can do."

He proceeded to give his best shot at explaining the opportunity as he saw it. They would take inspiration from natural processes, much as the other Wright brothers did when they engineered a way to fly after studying birds. "A tree can capture carbon dioxide from the air. We know it can be done. We've got to figure out how to do it." After talking nonstop for fifteen minutes, Allen was relieved when Klaus jumped in to share his insight.

"What do you think?" Comer asked his adviser after the presentation of ideas was over.

"I don't think this is venture capital," replied the skeptical adviser. "This is adventure capital."

Undeterred, Comer agreed to contribute $5 million and gave them a three-year timeline to produce a marketable prototype.

The Wright brothers boarded a plane back to Arizona to begin working on an air capture device. "Probably for the first forty-five minutes we just stared at the back of the seat in front of us," said Burt. "And then Allen said, 'Oh my God, what have I done? How can we pull this off?'" Allen opted for the vaguest possible name for the company, Global Research Technologies, as he was still uncertain what he was actually being tasked with. He rented warehouse space in Tucson, bought a pickup truck and started buying bits and pieces from junkyards and surplus auctions that looked interesting. Then he started asking basic questions: *How does water flow? What does water flow over? What does it do when you spray it in the air? What does air do? How do you mix the two? Do you want to mix the two? What shape do you want? What am I doing?*

He put a team together and, along with Klaus when he wasn't teaching at Columbia, they began conducting experiments. They believed that to find an economically feasible solution, a critical design feature would be passive air capture: letting wind carry air past the sorbent—which could be a liquid or solid—as opposed to using fans to move the air. They began using sodium hydroxide as a sorbent. As air passed by the sodium hydroxide solution it would react with the carbon dioxide, producing sodium carbonate. They tried spraying sodium hydroxide on hanging plastic strips and flowing it over stacked plastic wheels. They even tried it on venetian blinds from Target. The best approach from these early experiments was hanging shower curtain–like plastic sheets that were sandblasted to roughen the surface and spaced closely so that air could flow through—and react with sodium hydroxide cascading down—but not so close as to impede air flow.

There were drawbacks to working with sodium hydroxide, though. In a long-term application it would require expensive materials, given its highly caustic nature. But the greatest challenge was separating carbon dioxide from the sodium hydroxide once it had formed sodium carbonate. At Biosphere 2, calcium hydroxide was brought in to react with the sodium carbonate to strip the sodium hydroxide from the carbon dioxide and produce calcium carbonate, which is limestone. For the scale of Biosphere 2, it was feasible to store that limestone in drums. For a planetary scale, such a setup wouldn't be economic. A self-sustaining system was needed that would recycle the chemical sorbent and spit out carbon dioxide as a pure product. With sodium hydroxide as the sorbent, that could be achieved by heating the sodium carbonate to separate the carbon dioxide, but the heat needed would require lots of energy because of the strong bond created.

The team decided to try testing solid sorbents instead. They wanted to find something that would have uptake comparable to a sodium hydroxide solution but would also allow for easy separation of the carbon dioxide and reuse of the sorbent. One solid they stumbled upon was a membrane used for water purification that came in thin sheets. In their experiments, they discovered that the membrane had some unexpected properties. Wanting to better understand its capabilities, they loaded up a piece with carbon dioxide, rinsed it in water and left it overnight in a sealed jar.

A long list of experiments was lined up when they returned to the lab the next morning, but before they got underway, Klaus wanted to check on the overnight experiment. As expected, the carbon dioxide concentration in the jar was higher than the outside air. He opened the jar and let in fresh air then closed it again. The theory was that the membrane would release carbon dioxide and they would see an increase in the carbon dioxide level in the jar, back toward the level it was before the jar was opened. But instead of the carbon dioxide concentration in the jar going up from roughly 400 parts per million after being exposed, it started going down. "That was the point when Klaus said, 'Rip up the experiment list. There's something going on here that we don't understand. Until we figure this out, we're not doing anything else,'" said Allen. The only thing that had changed was the relative humidity inside the jar, from 100 percent when the membrane was soaked in water to nearly zero after being exposed to the dry desert air of Tucson. They were beginning to understand a moisture swing phenomenon, an ability of certain sorbents to alternate between off-gassing carbon dioxide when wet and uptaking it when dry.

They had found what they believed was a winning approach—a sorbent that could capture and release carbon dioxide without needing energy beyond that required to change moisture levels. They would just need a device in which to embed it. For that, they enlisted the help of an artist who drew up plans for air extraction prototypes. The artist's rendering of a small air capture farm, which resembled a shrunken downtown of office buildings, was included in a Columbia University press release, "First Successful Demonstration of Carbon Dioxide Air Capture Technology Achieved," issued on behalf of Global Research Technology. "The GRT's demonstration could have far-reaching consequences for the battle to reduce greenhouse gas levels," stated the press release. "The technology shows, for the first time, that carbon dioxide emissions from vehicles on the streets of Bangkok could be removed from the atmosphere by devices located in Iceland."

It still wasn't enough to convince Howard Herzog and other scientists that direct air capture might be economically feasible. "People want to believe in DAC because it is a simple concept that solves so many problems," said Herzog. He has used an analogy of separating marbles of different colors. To separate carbon dioxide from the smokestack of a power plant or industrial facility equates to removing roughly 400 red marbles from a bowl of 4,000. For direct air capture, it's akin to removing 400 red marbles from a bowl of one million. The number of marbles varies depending on whether the smokestack is attached to a less concentrated amount of carbon dioxide like a gas-fired plant or a more concentrated source like a cement plant.

Using the concentration from Herzog's marbles analogy, air capture requires processing 250 times more air, which is enough to overwhelm the economics, he argued.

To defend his perspective, he has pointed to a plot first developed in the 1950s called the Sherwood Plot. It shows that when something is extracted, the relationship between its concentration within whatever it's being extracted from and the cost of doing so is typically linear. If copper makes up 1 percent of the ore it is extracted from, it costs one dollar per pound to extract it. If gold makes up a thousandth of 1 percent, it costs nearly $1,000 per pound to extract it. Based on Herzog's estimate of what capturing carbon dioxide from a smokestack would cost, he extrapolated that direct air capture would cost on the order of $1,000 per ton of carbon dioxide. Klaus, however, insisted that direct air capture is not necessarily beholden to the Sherwood Plot. Applying the laws of thermodynamics led him to conclude that the energy cost of direct air capture would indeed be greater than that of capturing from a smokestack, "but the difference is quite small."

Wally Broecker wrote a research paper that attempted to arbitrate the matter, comparing a cost estimate from Klaus that it might cost $100 per ton with Herzog's $1,000 per ton. "If Lackner is correct, then, for example, the cost of retrieving the CO_2 produced by automobiles would add less than one dollar per gallon to the price of gasoline. This would certainly be affordable," wrote Broecker. "But if the [Herzog] estimate were to turn out to be the correct one, then it would add ten dollars a gallon to the price of gasoline." In attempting to validate which perspective was more accurate, Broecker was left to rely on the evidence he found most compelling: "As I happen to be a long-time colleague of Klaus Lackner, I am quite familiar with his views on this. I find him to be both a brilliant physicist and also a very thorough and honest man. He has spent much of the last thirteen years working on this problem. For this reason, I assign a much higher likelihood to his cost estimate than to those made by others."

For Klaus and Allen to prove up their concept on a larger scale, more money would be needed. With Gary Comer's death in 2006, his estate moved on to other climate initiatives. A Boston-based investment fund was interested, but the same morning they sat down to sign paperwork, Lehman Brothers shut its doors. The ensuing financial meltdown dampened any other potential sources of financing, and Global Research Technologies was forced to shut down. "The world wasn't ready," said Allen.

THERMOSTATIC AMBITIONS

KLAUS LACKNER AND Allen Wright were not the first to envision forests of artificial trees cleaning up the atmosphere. A Columbia University colleague named Peter Eisenberger had conceived of something similar in the 1980s: photosynthesis-mimicking contraptions that harvested carbon dioxide from the atmosphere and, using sunlight for energy, combined it with hydrogen to produce clean fuels.

Peter was trained as a physicist and began working at Bell Laboratories in 1968, a research and scientific company owned by AT&T. In the 1970s, oil supply concerns prompted Exxon to establish a similar research laboratory to catalyze what they believed at the time would be a necessary transition from being an oil company to an energy company. Exxon hired hundreds of PhD graduates, including Peter in 1981 as a director of their Physical Sciences Laboratory, and invested $250 million in creating the lab. Peter's vision while there was that energy could be produced by learning from nature's approach, perfected over many millions of years. Figuring out how to do it held the potential of creating an entirely new, sustainable industrial ecology.

But he was unable to act on his vision before Saudi Arabia declared in 1985 that it would ramp up oil production to reclaim market share; any post-oil planning was subsequently cast aside as oil prices returned to lower levels. Instead, Exxon shifted focus to undermining climate change mitigation, pursuing a subtle and systematic communications campaign to that end throughout the 1990s and 2000s.

In 1989, Peter moved on to be a professor at Princeton University, where he helped establish a research institute that brought together different disciplines to develop materials inspired by nature (known as biomimetic materials). Princeton was also home to a supercomputer that was used to model the behavior of the atmosphere and oceans. The first-of-a-kind modeling generated by that computer provided Peter with an introduction to climate change. In 1995, he was lured to Columbia University to head up the new Earth Institute, which would take an interdisciplinary approach to studying sustainable development. The guiding principle for the institute was that science and technology could be applied to not only preserve the natural systems that support all life on Earth but also improve the conditions of the world's poor. In practice, it was difficult to shift scientific focus to pressing societal issues as many faculty felt their focus areas were being devalued and, in turn, they questioned Peter's leadership. "The Earth Institute is a great idea," said Wally Broecker. "It's just got to be done in the right way." One professor who was supportive of Peter and his push to shift focus to less traditional realms like the economics of climate change was Graciela Chichilnisky.

Graciela, who is on a quest to reverse climate change, is a child of Argentina. The splendor and vitality she experienced growing up in Buenos Aires in the 1950s left a strong impression on her. So did the many people she saw being left behind as the world embarked on a period of globalization. The plight of low-income Argentines was championed by Eva Perón, the First Lady of Argentina, who was given the title of the nation's "spiritual leader" when Graciela was eight years old. Perón had grown up in poverty herself before ending up on what became a grand stage, urging the *descamisados* or "shirtless ones" to revolt against the oligarchy. Graciela's parents had also experienced poverty after fleeing Russia's anti-Jewish pogroms and finding themselves destitute in Buenos Aires. Her father, Salomon Chichilnisky, eventually found a job as a dock worker to support his family and went on to become a neurology professor. He became one of Perón's doctors, helping her battle cervical cancer, which had made her so weak that she could only appear in public next to her husband, Juan Perón, with the help of a fur coat bolstered with plaster and wire. Perón's death at a young age deeply impacted both the country and Graciela. Despite the mistakes she believes the Peróns made, which contributed to Argentina not realizing its ambition of becoming a wealthy country, Graciela was instilled with a passion to stand up for basic human needs.

Juan Perón's government was overthrown by a military coup a few years after his wife's death. Two coups d'états later, in 1966, Graciela was a precocious high school student, interloping on university courses while also raising a child she had given birth to the year prior. She had started with philosophy and sociology courses, believing they would put her on a path to helping humankind, but came to favor the clarity of math, wondering if she could make more of a difference by immersing herself in the formulas and models underpinning economic theories. This particular military coup, however, resulted in an indoctrinating, anticommunist military dictatorship that targeted universities as perceived incubators of communist subversion. On a night known as The Night of the Long Batons, police were ordered to forcibly purge the national universities. Those inside faced the choice of inhaling tear gas or exiting through a gauntlet of swinging batons.

One of the bludgeoned was a visiting professor from MIT named Warren Ambrose, who received "seven or eight wallops" and broke his finger but was not seriously injured, according to reports from the U.S. embassy in Buenos Aires. He was relieved to be able to return to the academic freedom he enjoyed in the U.S., unlike many of the tenured professors who went into exile or joined private institutions to ride out a desolate academic milieu that lasted until democracy returned nearly two decades later, in 1983. Sensing that the students he taught would also be better off in the U.S., he brought six of them with him to complete doctoral programs at MIT. Graciela was one of them. She was accepted into the program and given a scholarship by the Ford Foundation despite not having even graduated from high school. Her newly appointed adviser still voiced concern that she was attempting an impossible feat, being a woman with a young child in a new country, speaking a new language and not having completed one university course. She managed, however, not only to complete her classes and graduate with a PhD in math but to get another PhD at Stanford in economics, assisted by a campus childcare center she helped establish along the way.

She has not slowed down since; her résumé now runs forty-three pages and is filled with faculty positions at Columbia University (where she has taught since 1977), board seats, advisory roles, editorial positions, scientific papers, books and awards. She believes that for women to be successful professionally, at least in the Ivy League context in which she's been immersed, they must turn negative responses into positive resources, or as she puts it, "dung into fertilizer." She also believes that the only genuine source of happiness is the feeling of being useful to others. To that end, much of her energy has been focused on attempting to address inequities she sees, from pushing the UN to adopt basic-needs criteria—food, housing, education—as a more appropriate way to gauge progress than GDP maximization, to taking Columbia University to court on multiple occasions to seek an end to gender discrimination against female faculty members.

Much of her work has focused on addressing inequities she saw in how the world's shared environmental assets—the land, air and sea, and potential economic resources therein—are valued. Initially, she focused on the value placed on land resources in the global south, which she believed were being overly discounted because property rights were often less well established than in the global north. With land often treated as common property in the south, longer-term stewardship frequently gave way to overexploitation by the north. One outcome of the exploitative relationship was that a low floor on resource prices was inhibiting innovation in using resources more sustainably. But convincing governments to hand over property rights to stewards more oriented to the long term was typically not a realistic undertaking. So, instead, she shifted her focus to more practical market-based solutions that would attach costs to using the "global commons" (the land, air and sea), like polluting the atmosphere, that were more reflective of true costs. That led her to focus on a more troubling inequity: the future costs of using the global commons were being discounted too heavily. Standard valuation approaches, which discount the value of a cost or benefit in the future in comparison to present-day costs and benefits, were creating a strong bias against future generations. The most impactful result related to climate change: the majority of costs were expected to occur far enough into the future that they could be sufficiently discounted to justify kicking the problem down the road.

By 2006, however, Graciela was losing patience with attempts to apply economic theory as a practical solution to climate change. It was also becoming clearer that merely cleaning up energy usage would be insufficient to address the problem; economic and political realities were standing in the way of decarbonizing quickly enough. From her perspective, the way the math was lining up, carbon dioxide would have to be removed from the atmosphere in massive quantities to avoid catastrophic impacts. Unbeknownst to her, Klaus Lackner and Allen Wright, both Columbia colleagues, had just invented a contraption that might be capable of such a feat. Her close colleague, Peter Eisenberger, had gotten to know Klaus, a fellow physicist, through Peter's efforts to establish the Earth Institute at Columbia University.

One of the courses Peter taught at that time was on closing the carbon cycle. Numerous natural processes—photosynthesis, respiration, weathering, fossilization—move carbon through different repositories in a closed cycle. The burning of fossil fuels opens the cycle by releasing carbon dioxide into the atmosphere, where it can't be absorbed quickly enough by the natural carbon cycle, thereby creating the greenhouse effect as the accumulation slows the rate at which heat radiated by the earth can escape into space. Peter was aware of Klaus's direct air capture concept, a possible option to help close the cycle, so he reached out to Klaus in 2007 to give guest lectures on the concept. As he listened to his guest talk about direct air capture in more detail to the students, he could barely contain his excitement; here was the missing component from his vision of mimicking photosynthesis to produce clean fuels. He eagerly approached Klaus after the lecture to brainstorm about using concentrated solar power to fuel the contraptions, produce electricity as a by-product and create an entirely new, sustainable industrial ecology. When Peter told Graciela about Klaus's prototype and his general ruminations on capturing carbon, she immediately saw the potential for direct air capture, especially if it could generate a profit. She suggested they too start a company to develop the technology.

It was left to Peter to devise the most efficient way to extract carbon dioxide from the air, so he holed up in a house overlooking the ocean in northern California to study what was known about capturing carbon dioxide. He landed on something similar to the catalytic converters used in cars that reduce the toxic pollutants exiting the tailpipe; except instead of using precious metals like the ones that make catalytic converters highly desired by thieves, he incorporated more inexpensive chemicals. The design was premised on using fans to draw air through a honeycomb structure coated with a chemical sorbent that binds with carbon dioxide while minimizing the energy required for moving the air through. The honeycomb cells rotate at frequent intervals and, after being exposed to low-temperature steam, release carbon dioxide into a collection system. Then the chemical sorbent is reused and the process repeats itself. Graciela envisioned not only commercializing a technology that could capture vast quantities of carbon dioxide while making a profit but also meeting people's basic needs in the process by using carbon dioxide to provide fresh water (through a process referred to as biodesalination, which uses photosynthetic bacteria) or locating direct air capture plants with the aim of improving the economies of surrounding communities.

Financing for their new venture came relatively easily thanks to one of their students being connected to a very wealthy family. Ben Bronfman was on his second attempt at completing a post-secondary education, his first having been derailed by aspirations of a musical career. He was inspired to pursue environmental studies at Columbia after watching *An Inconvenient Truth*. After sitting in on Klaus's guest lecture in Peter's class and learning of the entrepreneurial ambitions of his professors, whom he held in high regard, the young Bronfman introduced them to his father and grandfather, Edgar Bronfman Jr. and Sr., to see if they might be interested in investing in their startup. The patriarch of the family, Sam Bronfman, had started a distillery in Canada in 1924 and grew it into the Seagram beverage empire. The Bronfmans have since become well-known for their philanthropy and diverse investment holdings, and they agreed to invest in Peter and Graciela's business. Edgar Jr. became the executive chairman of the board, and Ben was brought on as a partner (Peter and Graciela remained in control of the patents and intellectual property). One of Edgar Jr.'s investment philosophies is that the size of the opportunity is always equal to the size of the problem. "The climate solutions that are created are going to create the largest asset class in the history of the world," he said.

Graciela became the CEO, and the company was named Global Thermostat, which reflected not only a vision for the technology to potentially influence global temperatures but also a tendency she has exhibited to aggrandize its potential.[2] "I will complete the job of reversing climate change, [and] turn Global Thermostat into a $1 billion company," she said in an interview. Such a monumental feat—which would presumably garner a much greater valuation—wasn't even sufficient; she went on in the same sentence to say she will also "revolutionize quantum theory using topology for the benefit of understanding what is time and who we really are." So far, progress has been slow. A few direct air capture pilot projects have been constructed, with a total capacity of a few thousand tons per year, but the company is a long way off from realizing its stated goal of removing thirty to forty billion tons of carbon dioxide per year by 2050.

Peter, whose official title is co-founder and board member, is more realistic about what will be required for direct air capture to reach a meaningful scale. He estimates that worldwide direct air capture capacity will have to increase at a faster rate than for any other technology in history, including the Chinese ramp-up in solar manufacturing that saw capacity double roughly every year. And he acknowledges that a single company is not capable of achieving such a feat alone. "So the real issue is to develop cooperation in the industrial sector, and across the planet, to focus on this threat," he said. "But recognizing that rather than its current portrayal as a cost to the economy, it's really an opportunity to grow the economy." Peter has long viewed that challenge

---

2   It wasn't Graciela Chichilnisky's first attempt at commercializing a new technology. The first company she started up, a telecommunications company called FITEL, facilitated electronic trading of financial securities in the 1980s. Ironically, one of her first hires was Jeff Bezos, who went on to found Amazon and become the world's richest person while paying relatively little tax, including two years in which he paid no federal income tax at all, personifying the inequity that Graciela set out to rectify.

through a practical lens, but he has since added a more emotional one, a change in perspective that he attributes to the COVID-19 pandemic. "The devastation it created in a global economy and among people and the lives it took really made me realize that this was child's play compared to what is going to happen if we have catastrophic climate change."

A result of Peter's reorientation is that he is putting greater personal emphasis on creating momentum to collaborate on building a new industrial ecology. That aspiration goes back to the artificial photosynthesis vision he had while working in Exxon's research lab in the 1980s: that energy production could become more of a disciple than a foe of nature. "A renewable energy and materials economy based on the inputs to production being the sun, CO_2 from the air and hydrogen from water is the new industrial ecology we would like to build together," he said. "Once you have CO_2 and hydrogen, the petrochemical industry knows how to convert that to almost anything you want." Examples of potential outputs are producing carbon fiber, a much stronger per weight alternative to steel and aluminum, or synthetic aggregate made from carbon dioxide and calcium sourced from waste as an alternative to concrete. "Many people say to me, 'Oh, Peter, that doesn't make sense because what you're doing to get that carbon is undoing the bond between carbon and oxygen, and that's the very bond that was formed during combustion. That's going to be so expensive.' I point out to them right away that the aluminum and steel industries together account for 10 percent of our energy usage because it's much more costly to separate iron from its ore and aluminum from its ore than it is to remove carbon from its oxygen."

The result is a positive feedback loop: carbon dioxide is removed from the air where it is harmful and used to make important products while stimulating job growth and economic development in the process. "That's exactly what nature does. After it produces energy for life in a leaf, it expires the CO_2 in the air.

When that CO_2 comes back in, it's divided into two things: part of it grows the tree trunk and structure, which is sequestering the carbon, and part of it is converted into sugar that is used as the basis for the energy of life," said Peter. "We're just going to mimic that." If one adopts that perspective, direct air capture has the advantage, similarly to a plant, of either allowing for the reuse of carbon to create new materials or the sequestration of it underground. Of course, mimicking nature on such a scale as to be meaningful is well beyond the capacity of one company. "If for some reason there's a problem with what we do, then take the next step and move on. It's a little bit like during World War II when Ford switched from making cars to making tanks," said Peter. "And Exxon switched to making rubber."

GLOBAL THERMOSTAT'S FIRST ATTEMPT at building a commercial-scale direct air capture facility came in 2018. On the outskirts of Huntsville, Alabama, a road called A Cleaner Way leads to a collection center for hazardous household waste. Close to the end of the road is an offshoot called Fresh Way, a nondescript, industrial cul-de-sac that hosts a produce distributor. The sunny street names would be an even better fit for hosting the world's first commercial direct air capture facility, which was the intended purpose of two green shipping containers with industrial fans mounted on top that were installed with the intention of supplying Coca-Cola with carbon dioxide for carbonating beverages. The project was designed to capture 4,000 tons per year, but it now sits idle. The company contracted to build it, Streamline Automation, sued Global Thermostat for $600,000 in unpaid bills, damages and interest. Global Thermostat in turn countersued Streamline, stating that Streamline was responsible for the facility not working, and began constructing a new facility in Colorado.

It was a blemish on Global Thermostat's track record that received some media attention in 2021. An article published in *Bloomberg* said that, according to insiders, the "pioneering, brilliant co-founder Graciela Chichilnisky" was holding back America's best hope to pull carbon dioxide directly from the air. "She kept driving people away," one former senior executive was quoted as having said. "No one could deal with her lack of ability to separate the personal from the professional." Another employee was quoted saying a lot of the team's effort was diverted to assisting with Graciela's work at Columbia, like grading papers, and personal matters, like filling out forms for medical appointments. Peter isn't interested in discussing internal matters, but he understands why observers might perceive a lack of progress. "In developing a new technology, failure is a learning experience," he said, referring to the Huntsville project. Had Global Thermostat stuck with the design used for a pilot project completed in 2010, he believes the company would have more installed capacity operating today. Instead, they chose to focus on optimizing the design to reduce costs, taking a step backward to take multiple steps forward.

Before the Huntsville facility was shut down, one of the world's largest oil and gas companies came for a tour. Exxon's VP of research and development, Vijay Swarup, was hired by Peter as a postdoctoral student in 1987 when Exxon ramped up its energy research laboratory. Now Swarup was reaching out to his former boss because he believed direct air capture could fill a key gap in Exxon's portfolio of technologies. Peter didn't have a fully operating facility to show Swarup and his colleagues, but nonetheless he told the group that carbon dioxide could be captured at the unbelievably low cost of $50 per ton. He says that was possible because of the internal research and development efforts that had led to performance improvements and cost efficiencies. Exxon spent another year evaluating the feasibility and scalability of Global Thermostat's technology before announcing that they were on board, and a joint development agreement was announced. Peter believes Exxon is genuinely interested in developing the technology. "You can argue whether that genuineness is because they recognize that it's very important to their business or because they also have grandchildren and they're worried about the future," said Peter. "I can tell you that for Vijay Swarup, it is both: he's concerned for his business, but he's also concerned as a human being for the future of this planet."

The Global Thermostat press release announcing the partnership stated that Exxon's financial and technological strength would enable the technology to be scaled up to capture a billion tons per year. While that would require tens of thousands of the company's direct air capture plants to be built around the world, the press release argued that the monumental ramp-up will be facilitated by following the same process that China undertook to lower the cost of producing solar panels. And incidentally, noted the press release, China received funding from the Kyoto Protocol's Clean Development Mechanism, a carbon-offset scheme authored by Graciela Chichilnisky herself. When pressed for more details about how such partnerships might lead to such a giant leap in capacity, Peter says that Global Thermostat will grant non-exclusive licenses to anyone interested in using their technology to build their own facilities and is working with suppliers, construction firms, equipment manufacturers and others in a mobilization effort. When Graciela was similarly asked to explain how a partnership like the one with Exxon might lead to scaling up to billions of tons of year in capacity, her response was more taciturn: "The answer is, I don't know. These are humongous firms. I'm not telling these friends what to do."

HARD SCRUBBING

THE GREEN RIVER BEGINS NEAR the peaks of Wyoming's highest mountains as meltwater relinquished by glaciers. As it tumbles through the Wind River Mountains, the river begins to swell before exiting the mountains on a meandering southward journey through a landscape of sagebrush and steppe. Despite the desert-like conditions of the surrounding basin, abundant wildlife follow the river's corridor. During winter months, extensive herds of pronghorn make lengthy migrations from the mountains along the Green River Basin to forage on the sagebrush exposed by a sparser snowpack than the terrain they vacated.

There is another attraction in the area that is much more inaccessible: carbon dioxide lying far below the surface. Despite the miles-thick barrier, and the unsafe overabundance in the air above, carbon dioxide here is still brought to the surface. Some of it is sold, then transported to another location and pumped back below the surface to be used for enhanced oil recovery, and some of it is released into the atmosphere as a by-product of ExxonMobil's LaBarge operation (other gases are produced alongside the carbon dioxide—methane, nitrogen, hydrogen sulfide, helium—but only helium generates a profit).

Upstream of where the LaBarge facility sits along the Green River are the river's headwaters, located in Wyoming's Bridger-Teton National Forest. Chris James was hiking alone in that pristine watershed one day in 2019 when he had an epiphany. During a recent family dinner, his son's questions, which had arisen from learning about climate change at school, had caused him to question whether he could continue compartmentalizing his personal values and investment philosophy. He was a conservationist and an energy investor—two increasingly incongruent pursuits. He wondered if perhaps there was a way to wake up in the morning and work on a mission that was consistent with his values.

Chris grew up in the coal mining town of Harrisburg, Illinois. He left after high school, made his way east to Wall Street and then, sensing a better opportunity on the west coast, headed to Silicon Valley. His timing was fortuitous, and he made more good calls than bad calls. He decided to diversify away from the tech sector when an opportunity arose to invest in a coal mine near his Harrisburg hometown. He liked the premise of helping the community by replacing some of the jobs lost when the area's high-sulfur coal mines were forced to shut down in the 1990s. What he didn't appreciate was how rapidly demand for coal was dropping as natural gas prices, driven to sustained lows by the advent of hydraulic fracturing, were propelling gas-fired power plants to outcompete coal-fired competition. It was a pivotal lesson in how change driven by innovation can be nonlinear.

As Chris hiked near the Green River headwaters, he wondered if reconciling his pursuits would not only bring more congruity to his life but also make better business sense. What if there were different benchmarks for judging investments' worthiness that could be better predictors of long-term success *and* make society better off overall? A 2017 paper by economists

Oliver Hart and Luigi Zingales had influenced Chris's ruminations. The authors questioned the widely accepted belief,
popularized by economist Milton Friedman in the 1970s, that
a company's officers and directors should prioritize making
money above all else, so long as they operate within the confines
of the applicable society's rules. If one continues with that logic,
then the matter of externalities, say the impact on the climate of
burning fossil fuels, should be dealt with by consumers or governments. That argument breaks down if consumers don't have
the means of offsetting the externalities of a company's operations, leaving governments with the responsibility for dealing
with externalities. But the creation of effective climate policy
has been hampered by slow-moving political processes that are
often exacerbated by the complexity of trying to align policies
with other governments.

In the absence of the conditions envisioned by Milton Friedman, like governments not ignoring externalities, the paper's
authors argue that the goals of companies cannot be totally separated from the goals of individuals and governments. So, instead,
shareholders should prioritize maximizing their own welfare, or
general well-being, as opposed to strictly trying to maximize
the market value of their investments. To that end, shareholders should be able to vote on the broad outlines of corporate
policy. That does not guarantee that value maximization won't
remain triumphant if the social preferences of shareholders are
too mismatched to be funneled into meaningful corporate policy.
Otherwise, shareholder welfare maximization is undeniably a
more worthwhile objective.

Chris was convinced that the best way to integrate his personal values with his investing activities was to become an activist
shareholder (he prefers the term *active owner*). Not only would
that allow him to wake up with one mission to focus on, but he
was increasingly convinced it was also the *best* value maximization approach. That was most readily apparent when it came to

climate action: all the research he did suggested that the costs of decarbonizing—renewables, electricity grid battery storage, electric vehicles—were dropping so quickly that adoption was set to explode.

Shortly after Chris's epiphany in the Wyoming woods, he was introduced to someone with activist investing experience, a hedge fund partner named Charlie Penner. Charlie had been involved in two recent activist campaigns, pressuring Apple to enhance parental control on devices and pushing McDonald's to offer plant-based burgers. He was trying to find partners to take on a much larger campaign. Many oil and gas companies were starting to take climate change more seriously, but one of the largest seemed to Charlie to be most adrift: Exxon.

Profits at Exxon had gone from the largest ever realized by a U.S. company in 2008 to steadily declining. When Charlie and Chris joined forces in 2020 at Chris's new "impact-focused investment firm," named Engine No. 1 after one of the oldest fire stations in San Francisco, Exxon was about to report a $22 billion loss for the year. At the same time, the company was spending $17 billion on capital expenditures, mostly earmarked for growing production. According to internal documents leaked to *Bloomberg*, it was part of a five-year, $210 billion investment plan targeted at increasing production from around the world. The plan would have resulted in the company's emissions growing by 17 percent over the same period. The company's total emissions, including those associated with burning the fossil fuels it produces, are roughly equivalent to the amount emitted by South Korea, a country of over fifty million people. The leaked plan would see the company add the equivalent of Greece's emissions.

WHEN THE MANAGEMENT of publicly traded companies puts forth a slate of directors for shareholders to vote on each year, it includes directors seeking reelection and any nominees to fill vacant spots. The preference at Exxon has been to nominate individuals for board seats who have experience as executives in other businesses, such as pharmaceuticals and insurance. Other candidates nominated by shareholders are considered if they are deemed to have requisite experience. In theory, that meant nominees for the 2021 director vote put forth by Engine No. 1, a new Exxon shareholder, would be given consideration if they met Exxon's prerequisites. Chris and Charlie started seeking nominees who would fit Exxon's criteria but who also recognized that the energy giant's business model would have to change for shareholder welfare to be maximized.

One of their recruits was a Finnish woman named Kaisa Hietala, who worked for a consulting firm called Gaia Consulting Oy. Before joining Gaia in 2019, she led the transformation of a Finnish refining business into the world's largest producer of renewable diesel. The makeover led to the renewable fuels division generating over 90 percent of the company's profits and a return on investment of over 500 percent for shareholders, and it was named one of the top twenty business transformations of the previous decade by *Harvard Business Review*. Another Engine No. 1 nominee, Andy Karsner, was appointed by President George W. Bush as the assistant secretary for the Office of Energy Efficiency and Renewable Energy and by President Obama to the National Petroleum Council. Anders Runevad, another nominee, had been the chief executive officer of Vestas, the world's largest wind turbine manufacturer. During his six-year tenure, the company's stock returned a total of 480 percent to shareholders.

The most orthodox recruit—putting aside Exxon's policy of hiring executives from other industries—was an oil executive named Greg Goff. He had presided over a refining and marketing company called Andeavor from 2012 to 2018, the shares of which generated a return of over 1,200 percent for shareholders over that period. He saw the business case for industry change and was viewed as important for establishing credibility by rounding out the slate of nominees with a proven oil industry executive. After a few conversations, Chris and Charlie flew down to Texas to meet with Goff in person, hoping to entice him to join their crusade. He picked them up in a truck stocked with shotguns intended for shooting clay birds. They didn't leave the hunting lodge. Instead, they spent the day discussing the future of the energy industry, a conversation that extended late into the night. Goff wasn't ready to entertain the possibility of a world without fossil fuels, but he conceded that something had to give.

Engine No. 1's pitch to entice shareholders to vote for the four nominees at the 2021 annual general meeting, entitled "Reenergize ExxonMobil," began with a comparison of Exxon's return to shareholders over the previous five years of negative 33 percent to a positive 12 percent return for its peer group. It also showed that the company was becoming buried under an increasingly burdensome debt load, the largest in the company's history. But the biggest concern for shareholders, argued the presentation, was the potential for even worse future performance: "A refusal to accept that fossil fuel demand may decline in decades to come has led to a failure to take even initial steps towards evolution, and to obfuscating rather than addressing long-term business risk," read the presentation. It also drew on what Chris had learned from his brief foray into coal mining. A graph showed the U.S. Energy Information Administration's biannual forecasts for coal production beginning in 2010: every estimate was roughly 10 percent lower than the previous one so that forecasted production was about half of what it was a decade earlier.

Global Thermostat was also highlighted. The presentation pointed to the $15 million Exxon invested to kickstart the partnership and how Exxon had promoted it in commercials and social media posts. Exxon hadn't gone as far as Global Thermostat did in suggesting that the partnership, by leveraging the larger company's financial and technological strength, would lead to pulling a billion tons of carbon dioxide a year from the atmosphere, eventually growing to forty billion. Nonetheless, Engine No. 1 characterized Exxon's commitment as more emblematic of a marketing-centric scheme than a visionary business strategy. The pitch cited the *Bloomberg* article that questioned whether Graciela Chichilnisky's management approach was holding the company back: "Accounts suggest [Global Thermostat] has been stymied by setbacks and mismanagement since almost the very beginning and has made little progress in deployment over the past decade... Current and former staffers say it's unclear exactly what Exxon is doing with Global Thermostat besides advertising it heavily."

Overall, Engine No. 1 made a very compelling case for how to improve shareholder value, without even getting into shareholder welfare. Management did not agree, which set the stage for a battle. Exxon's nearly three million shareholders would be entitled to vote at the upcoming annual general meeting. Not all shareholders who planned to vote would attend in person, so it would be a proxy battle waged via third parties used to solicit votes. Exxon stated it would spend at least $35 million more than it normally would on the annual vote (Engine No. 1 speculates the figure was closer to $100 million more) as it inundated shareholders with over 130 filings arguing the company's position, along with a website, Twitter posts, blogs and employee forums to further resist the push from Engine No. 1. "Don't be deceived by a months-old hedge fund, Engine No. 1, that wants your Company to pursue a vague and undefined plan—which we believe will jeopardize our future and your dividend," wrote the company.

One of Exxon's arguments against electing the Engine No. 1 nominees was that they failed to meet the company's standards. The company's bylaws stipulate that board members have "demonstrated expertise in managing large, relatively complex organizations, and/or, in a professional or scientific capacity." Exxon also said the fund's four nominees did not meet the general board criteria of having served as CEOs of large public companies. Then, less than two weeks later, Exxon added a new board member who had been the CEO of a state-owned company, Malaysia-based Petronas, named Wan Zulkiflee. A month later two more nonconforming directors were added. None of the three individuals added to the board met the company's own criteria of having served as CEOs of large public companies; conversely, two of the Engine No. 1 nominees, whom Exxon would not even meet with, did.

Another of Exxon's arguments was that Engine No. 1 was critical of carbon capture, based on a quote from their investor presentation stating that "carbon capture is vaporware." The quote was misused; Engine No. 1 wasn't being critical of carbon capture, only its apparent use as a smokescreen by Exxon. The vaporware comment referred specifically to the announcement of a $100 billion carbon capture project that appeared out of nowhere when Engine No. 1 came along. The project, which would capture fifty million tons per year of carbon dioxide from refineries and petrochemical plants that line the Houston Ship Channel, was announced along with a request to the government for either tax breaks to fund it or a price on carbon to create a market for the emissions reductions. The Engine No. 1 team was skeptical about Exxon's sincerity in pushing for a carbon price. The presentation included a quote from the Union of Concerned Scientists stating that Exxon had consistently funded members of Congress who oppose putting a price on carbon.

A company's emissions are typically categorized as Scope 1, 2 or 3. Scope 1 emissions are from owned or controlled sources, mainly from fuel combusted in the process of producing something. Scope 2 emissions are from purchased electricity, heat, steam and cooling used by a company. Scope 3 emissions are all the other emissions that are indirectly related to a company's business, such as from other goods and services in a company's supply chain or, most significantly for oil and gas companies, from the use of sold products. In 2021, Exxon had disclosed their Scope 3 emissions for the first time: an estimated 650 million metric tons for the prior year, which dwarfed the 112 million metric tons of Scope 1 and 2 emissions combined. The problem for oil and gas companies is that carbon capture and sequestration can be used to reduce Scope 1 and 2 emissions but not Scope 3, which in Exxon's case totaled 85 percent of emissions. It's better than doing nothing, but carbon capture on its own would only amount to reducing a small fraction of the company's climate impacts.

Other parts of Exxon's counteroffensive seemed to acknowledge the risk posed to their business that Engine No. 1 was shining a spotlight on. For example, the creation of a new business group within Exxon was announced, Low Carbon Solutions, that would invest in carbon capture and storage opportunities and add other low-carbon technologies as they matured to commercialization. Perhaps Engine No. 1's proxy battle had encouraged Exxon to accelerate towards reducing emissions, but other actions made it difficult to believe they fully embraced the forthcoming transition. "This is the same company that for years has refused to take even gradual material steps towards being better positioned for the long-term in a decarbonizing world, and the depth of its efforts to fight off having just one-third of its Board possess the relevant experience to help it do so speaks volumes about ExxonMobil's future intentions" was Engine No. 1's closing argument to shareholders.

The day of Exxon's shareholder meeting, May 26, 2021, arrived with news that activists were driving another large oil and gas company to decarbonize more rapidly. An environmental group, alongside over 17,000 Dutch citizens, had succeeded in convincing a Netherlands court to force Shell to reduce carbon dioxide emissions by 45 percent by 2030, relative to 2019 levels. At the outset of the Exxon shareholder meeting, Charlie Penner was given the opportunity to speak on behalf of Engine No. 1:

The good news, we believe, is that no matter the outcome of today's vote, change is coming. Since this campaign started ExxonMobil has promised, in a number of different ways, that it will stop fighting and start facing the future. And these are promises it cannot easily walk away from, at least not if shareholders hold the board's feet to the fire. And if there's one thing that we've learned from this campaign, it's that *that* is what it will take... But we also learned that change can happen anywhere. It will always be a long shot, but it will always be worth it.

Charlie's remarks were followed by an opportunity for shareholders to present proposals, the majority of which called for changes that would make it easier to hold management and directors accountable for climate action. After forty-four minutes, during which all the shareholder proposals were rejected, management asked for an unprecedented hour-long recess. Voting was already underway, and management and directors were becoming increasingly concerned that it was not going as hoped. They spent the break phoning the largest shareholders.

When preliminary results were announced at the end of the meeting, two Engine No. 1 nominees had been elected, Greg Goff and Kaisa Hietala, while it was too close to call for Andy Karsner. Anders Runevad, the former Vestas CEO, who was ranked the fourteenth-best CEO in the world by the *Harvard Business Review* when he stepped down in 2019, was not elected. "That's okay," Runevad told his Engine No. 1 allies. "I thought it was a long shot for us to win any seats, but I thought it was worth it." When the final results were announced a month later, Andy Karsner was also in. Most of the support for the Engine No. 1 nominees came from institutional investors. Out were three directors, including the recently appointed Wan Zulkiflee, who in an interview with *Bloomberg* the previous year had questioned whether there was much need to reevaluate oil and gas business models.

"Energy transition, as many call it, is just an additional energy requirement, instead of a transition," he said. "Oil and gas will still play a major role, but will be complemented by other forms of energy."

A year prior, CEO Darren Woods had expressed similar skepticism, arguing that other oil and gas majors' pledges to reach net zero emissions by 2050 amounted to beauty contests. His rebuttal to Engine No. 1 the only time they did speak directly, in a call held shortly after the proxy battle was kicked off, was that investing in renewable energy wouldn't leverage Exxon's strengths. Rather, argued Woods, technological breakthroughs would be required in areas such as carbon capture, alternative transport fuels and rethinking industrial processes. It was the same argument made by Engine No. 1—the difference between the warring camps seemed to come down to Exxon needing greater assurance about what the world will look like in the future before truly shaking up their business model.

"Hey, Charlie, do you know how anybody is going to meet the 2050 goal today?" asked Woods on the call. "Have you asked any CEOs who have committed to that?"

"Do you know how you're going to fulfill your business plan without burning down the planet?" replied Charlie.

"If all it takes is aspiration, we support that ambition."

"Have you ever accomplished anything that, when you started, you didn't know how you were going to finish?"

IN 2022, EXXON ANNOUNCED that the LaBarge facility—which sits adjacent to the Green River in Wyoming and pulls carbon dioxide from underground into the atmosphere—is expected to emit roughly one million fewer tons of carbon dioxide per year by 2025. "By expanding carbon capture and storage at LaBarge, we can reduce emissions from our operations and continue to demonstrate the large-scale capability for carbon capture and storage to address emissions from vital sectors of the global economy," said Joe Blommaert, president of Exxon's Low Carbon Solutions, the venture launched during Engine No. 1's campaign and tasked with commercializing Global Thermostat's direct air capture technology. The only thing Blommaert said publicly about direct air capture, before retiring in May 2022, was that research was ongoing.

TWO OPTIONS

A COMMON FEATURE OF MANY recently published books on climate change is a reference to a scientist named David Keith, usually in the context of solar geoengineering. David, a professor of applied physics and public policy at Harvard University, is one of the leading proponents of advancing studies of the eleventh-hour approach to cooling the planet, which entails the temporary placement of extremely fine particles in the upper atmosphere that would curtail the amount of solar radiance hitting the earth. It's not a recourse he thinks the world will want to rely on; it's a stopgap measure until carbon removal can mop up excess carbon dioxide after the world has (hopefully) decarbonized.

The particles that would be released into the upper atmosphere would be very, very small—about a thousand times smaller than the width of a human hair—and are referred to as aerosols because they stay suspended in the stratosphere about twenty kilometers above the earth's surface, but only for about two years. A million and a half to two million tons of sulfur dispersed from modified aircraft would suffice. David and other scientists calculate that would offset about half of the planet's warming and allow carbon removal to catch up before the program would gradually be wound down and the remaining suspended sulfur particles would fall back to earth as air pollution.

"Your obvious reaction should be: 'What? That's crazy!'" says David.

There are some considerations that may help allay concerns with the concept's apparent brazenness. For one, sulfur is already constantly entering the atmosphere. One natural source is phytoplankton, the typically microscopic plants found at the surface of the oceans and the bottom of the food chain. James Lovelock, whose Mars life-detection work led to developing the Gaia Hypothesis for how life on Earth is maintained by self-regulating systems, was intrigued at the time by the possibility that phytoplankton might play an integral role in the earth's natural sulfur cycle. After verifying that the algal organisms do indeed emit sulfur by studying samples found on a beach near his summer cottage in Ireland in 1971, Lovelock boarded a British research ship traveling to Antarctica to test the theory further along the way. Sulfur, or dimethyl sulfide to be more precise, was found in every sample collected—it's partly what gives the oceans their distinct smell—and some of that sulfur ends up as sulfate aerosols in the atmosphere. Those aerosols not only scatter sunlight, dimming what reaches the earth's surface, but also provide nuclei for water vapor to adhere to, thereby stimulating the creation of clouds that also shade the planet. As part of the sulfur cycle, some of that sulfur makes its way to terrestrial areas as rain and is an important nutrient for plants and animals.

Another, more volatile source of sulfur is volcanoes. The 1991 eruption of Mount Pinatubo in the Philippines was the largest volcanic eruption in the last century. Billions of tons of rock were sent into the sky by the gases released from the magma chamber. Twenty million tons of that gas was sulfur dioxide— it was, in effect, a very large solar geoengineering experiment.

The eruption was both large enough to impact global temperatures and recent enough that the effects could be accurately monitored. The sulfur dioxide that made its way to the stratosphere created an aerosol mist that quickly spread around the world and precipitated a drop in global temperatures of about 0.5 degrees Celsius from 1991 to 1993. It helped validate modeling of the effects of stratospheric aerosols (it also further confirmed modeling of the effects of carbon dioxide emissions) and provided scientists with useful information in studying the impacts solar geoengineering might have on the ozone layer, regional hydrological cycles, plant growth and other ecosystem responses.

But the greatest experiment with putting sulfate aerosols into the atmosphere is the many million more tons spewed each year from tailpipes, smokestacks and other flues. Global sulfur dioxide emissions have steadily declined since the 1980s, thanks to pollution controls implemented because of concerns with acid rain and other negative impacts on human health and the environment. But fossil fuel combustion is still the predominant source of atmospheric sulfur. The cooling effect of particles tied to remaining sulfur dioxide emissions, which end up in the lower atmosphere, is not nearly as strong as stratospherically placed particles. There are also significantly more variables at play in the lower atmosphere compared to the stratosphere. As a result, to what extent solar geoengineering would be replacing the planet-cooling effect of decreasing sulfur dioxide pollution, compared to offsetting the planet-warming effect of increasing carbon dioxide pollution, is still an open question.

One way to mitigate potential risks associated with solar geoengineering, according to David and other scientists, is to inject aerosols gradually into the stratosphere. A slow ramp-up would allow for plenty of monitoring along the way, including any needed adjustments, and the program could be effectively unwound in only two years, the approximate time it would take for all the aerosols to return to earth. And if the program is designed to only offset half the human influence, the likelihood of more undesirable effects would be reduced, like the possibility of lower global precipitation.

There is also a mitigating factor that represents a double-edged sword: it's cheap, or at least relatively so. The upfront costs to develop the aircraft have been estimated at a few billion dollars, after which it would cost roughly a billion dollars per year to run the program. That makes it a tiny fraction of the cost of other climate action. There are a lot of governments around the world that could afford to run a program like that, argues David, even if they were acting on their own. That should bolster the argument for extensive testing: Why not gain the benefit of having as much knowledge as possible before one or more countries undertake a rogue effort on their own? If the science is understood through years of rigorous research, at least such a program could be carried out in a way that would maximize positives and minimize negatives globally.

Many of the concerns with solar geoengineering voiced by critics are also shared by David. One of the most vexing is the impossibility of ever fully understanding the potential impacts before attempting it on a large enough scale to be impactful. What if one brutally ugly technical fix leads to unforeseen effects that require another fix, and so on, to the point that it becomes impossible to backtrack? How would those decisions even be made, especially if there will be winners and losers, when it's so difficult for countries to agree on much more straightforward decisions? What if one or more countries disagree with the approach taken by others and retaliate or initiate a competing solar geoengineering program?

Nonetheless, as a vocal champion for increased research, David has served as a lightning rod for discontentment that we are now contemplating such a desperate climate solution. Naomi Klein, in her book *This Changes Everything*, suggests that David is influenced by having a vested interest: "Many of the most aggressive advocates of geoengineering research are associated with planet-hacking start-ups." Yet there are no startups associated with solar geoengineering, and David is campaigning most vocally for other people to complete solar geoengineering research and lessen the burden on the existing small number of researchers. (At the same time, he's not aggressively promoting direct air capture, where he does have a vested financial interest and which is more waste management than planet hacking.)

Another critic, Michael Mann, is a well-known scientist who's battle-hardened from many run-ins with climate deniers. He puts solar geoengineering proponents in a doomist camp at the other end of the spectrum from deniers, arguing they're too quick to believe more conventional climate mitigation is no longer sufficient. He also suggests that David might suffer from "some degree of hubris" in leaping from modeling the effects of solar geoengineering to making conclusions about actual impacts and that he has moved too far along the path from dispassionate inquiry to advocacy. David holds a healthy amount of skepticism with respect to decision-makers' willingness to look out for the long-term health of the planet, but he's also buoyed by recent progress made towards cutting emissions and is motivated to see what he refers to as a "brutally ugly technical fix" researched further because of its potential to reduce suffering. Prudent contingency planning, it would seem, as opposed to the idea that, having thrown in the towel on other options, he's recklessly promoting the use of solar geoengineering. His long-time colleague, Ken Caldeira, has stressed that advocating for solar geoengineering research does not imply that one is advocating for its deployment. On the contrary, it puts scientists in the uncomfortable position of becoming experts on something they dislike. "It's a clear case where the people who developed the models didn't want to conclude that solar geoengineering would work, but that's what the models seem to say," he said on a podcast co-hosted by David, *Energy vs Climate*. Another solar geoengineering researcher who has worked with David, Jason Blackstock, says that scientists did not become experts on the subject "because of hubris, but because of fear."

In the opening to David's book on solar geoengineering, *A Case for Climate Engineering*, he writes, "Many people feel a visceral sense of repugnance on first hearing about geoengineering. That intuitive revulsion strikes me as healthy." He thinks the best remedy is to better understand the risk/reward tradeoff through testing, but many people can't move beyond that initial reaction. A 2021 test in Sweden would have been the first to deploy reflective particles to the stratosphere. A balloon would release a very small amount of materials, weighing less than a kilogram, to better understand how aerosols form in the stratosphere. But it was canceled by the Swedish space agency after opposition emerged. "Too dangerous to ever be used," argued the president of the Swedish Society for Nature Conservation.

That is the impasse that David and other advocates for more solar geoengineering research find themselves in: more testing is necessary to counter potentially exaggerated claims, but those same claims are preventing the experiments needed to test any theories. He is not pushing to start implementing solar geoengineering now; he's advocating for more time for testing—ideally more than a few years' worth—to better understand the impacts so that more educated decisions can be made in the future about if and how to deploy it, along with emissions abatement and carbon removal, in the hopes of making the planet as hospitable as possible.

GROWING UP, DAVID WAS EXPOSED to a mélange of ecology, physics and engineering. His father, a British-born, U.S.-educated field biologist, worked for the Canadian Wildlife Service regulating the use of DDT. They were joined on wilderness excursions by David's stepmother, who was also a biologist. Interesting house guests further piqued an interest in the natural world: an uncle who founded the American Birding Association and set several birding records, a colleague of his father who studied polar bears and spoke glowingly about life in the Arctic. Spending time in nature was preferable to studying—he hiked the Appalachian Trail on his own in high school—as reading and writing did not come naturally at the time due to challenges overcoming dyslexia. His stepmom thought he might be interested in the physics projects that a friend's husband was working on, so she lined up a visit to the research laboratory where Paul Corkum, a now well-known physicist in scientific circles, was operating. David asked a bunch of questions and was offered a summer job even though high school students were a rarity in the lab. "It was one of the coolest laser labs around—it had a huge impact on me. Paul is a real friend and mentor. He really got me to think seriously about physics," said David, who returned for several more summers in the lab. "There were all these amazing machinists and electrical engineers and so on. I was able to do amazing stuff with minimal supervision in a way that wouldn't be possible now that Paul's very famous."

David studied physics and philosophy at university but wasn't sure if those fields were something he wanted to pursue further as he feared they weren't relevant enough to everyday life. Instead, he tested other passions: climbing in the mountains to see if he wanted to become a mountain guide and making a detour to the high Arctic to assist with a walrus study and test his interest in biology. The pull of physics eventually won out, and he landed at MIT studying experimental physics, partly because of a recommendation from Paul Corkum and partly because he had won a national physics competition. While there, he fell in with a group of graduate students studying climate change and ended up researching geoengineering, a nascent field of study at the time that encompassed potential interventions in the climate system. "It was a pretty unusual group. We were kind of ahead of our professors," said David. "I started working on this topic simply because it was a topic somebody brought up and nobody was working on it."

He convinced a professor of climate policy at Carnegie Mellon University to collaborate on modeling the impacts of geoengineering options. In 1992, they published a paper that separated climate action into three categories: abatement, adaptation and geoengineering. "Doubt about the prospects for cooperative abatement of global GHG [greenhouse gas] emissions is a pragmatic reason to consider geoengineering, whose implementation requires fewer cooperating actors than abatement," wrote the prescient authors. Several different geoengineering approaches were considered—ocean fertilization, tree planting, solar shields, solar geoengineering—that aimed to either remove carbon dioxide from the atmosphere or reflect sunlight back into space. As a result of the paper, David was invited to join Carnegie Mellon to complete climate research in a department that combined both engineering and public policy.

Crude climate fixes were far off most radars at the time, and so David focused on more pressing matters, like trying to better understand the uncertainty behind key variables in climate modeling. A paper he co-authored in 1995, one of the most widely cited of the 200-plus papers he's helped write, tested scientists' consensus on estimating a key variable: climate sensitivity, the temperature response to a doubling of the atmospheric carbon dioxide concentration.

A century previously, a Swedish scientist named Svante Arrhenius made the first rigorous attempt at estimating climate sensitivity. Arrhenius, who began his career as a physics professor, became convinced that his focus area of physical chemistry no longer had potential for producing the most enlightening research. So he shifted his focus to what was known as cosmic physics, the interrelatedness of terrestrial, atmospheric and cosmic events; it was viewed not as much a recognized discipline but more of an intellectual hobby drawing on different fields of study. One of the hotly debated topics at the end of the 19th century— at least among those with an interest in cosmic physics—was the cause of Earth's previous ice ages. Arrhenius hypothesized that variations in carbon dioxide levels were likely the cause: to drive the variations experienced over millions of years would have required a change in carbon dioxide levels on the order of 50 percent. But he needed to demonstrate his theory quantitatively, so he spent most of 1895 completing tens of thousands of tedious calculations. He finished feeling annoyed that "so trifling a matter has cost me a full year," forecasting that, given the rate at which coal was being burned at the time, it would take 3,000 years for carbon dioxide levels to increase by 50 percent (it has actually taken a little over one century). His estimate for a temperature response from a doubling of carbon dioxide concentration was 4 degrees Celsius.[3]

David's research, which included interviewing leading climate scientists, found a high degree of consensus on estimating climate sensitivity (less so for other, more specific impacts). The mean estimates of climate sensitivity with a doubling of carbon dioxide concentration, from the pre-industrial 280 parts per million to 560 ppm, ranged from 2.2 to 3.3 degrees Celsius. But David and his co-author also found that reducing uncertainty further would be a slow process, pushing the Intergovernmental Panel on Climate Change (IPCC) to reconsider its view that it might only take ten to fifteen years to substantially do so. (The IPCC's best guess with respect to climate sensitivity has moved from 2.5 to 3.0 degrees Celsius.) David worked on understanding uncertainties for many other climate applications and how experts navigate around opaque variables in reaching conclusions.

In 2004, he was enticed to join the University of Calgary, which was embarking on an ambitious effort to establish a major climate and energy policy center, dubbed the Institute for Sustainable Energy, Environment and Economy. One of the attractions for David, who was appointed head of the institute,

---

3   His original estimate of 5–6 degrees Celsius was revised in 1906 to 4 degrees.

was the opportunity to embed himself in the energy sector—what better way to understand the industry he believed would need to change? Unfortunately, it didn't turn out that way. "There were a lot of good intentions. A number of individuals were serious about doing something about climate," said David. The problem arose with managing conflicts of interest at a university funded largely by one industry. "What disturbed me the most was that a university think-tank refused to do what a university should do: bring in diverse views and have strong debate. The government and industry didn't want that."

David contrasts his experience at the University of Calgary to Carnegie Mellon University, which acted as a neutral convener looking for creative ways to stimulate more interaction, and better analysis and outcomes as a result. As an example, he recalls annual meetings held with otherwise warring interest groups: "The formal part was showing off a bunch of academic papers. Who really cares? What really happens is that's the place where we get some of the people who are at the top of industry and some people from the top of the enviros who are suing them and some members of Congress and congressional staffers into one room to listen to a bunch of papers—and talk. And I think that really helps to advance policy compromise in a serious way."

With a constrained ability to balance competing interests from the inside, morale among the University of Calgary faculty dropped and recently recruited professors began to leave. David felt hamstrung in trying to build the institute, and so he devoted more time to research interests. In 2006, he was introduced to an early Microsoft employee named Jabe Blumenthal, who designed the first version of Excel and subsequently left to become a high school teacher, while also becoming a staunch advocate for climate action. Blumenthal maintained a relationship with Bill Gates and began influencing the Microsoft founder's burgeoning interest in addressing climate change. When Gates transitioned to a part-time role at Microsoft to free up time for new roles,

Blumenthal thought David might be able to help. David was invited to an initial session and brought along Ken Caldeira. One climate tutorial turned into about twenty-five over the course of several years, often with other experts invited by David and Ken to speak about specific topics. In advance of each session, Gates was provided with reading material ranging from hundreds to over a thousand pages. "He had clearly done the reading for every single one," said David. "And there was 100 percent focus to the point where there would be almost zero small talk over the course of the five-hour meetings." It was the kind of deep dive that David wishes bigger governments would do more of.

One outcome of the meetings was that Gates started giving his two climate advisers a million and a half dollars per year to fund climate research. "That gave me leeway to take risks," said David. He had been interested in carbon removal ever since he published the 1992 paper looking at geoengineering options. With some inspiration from Klaus Lackner, he had begun looking specifically into direct air capture at Carnegie Mellon, starting with an analysis of expected costs. That morphed into working directly with Klaus to determine what it would cost to build out Klaus's concept. "It's hard to think of another field that is so much the product of a single person's thinking and advocacy," says David. "Klaus was pivotal in making the argument that direct air capture could be developed at a scale relevant to the carbon-climate problem." Now, with Gates's money and more time, David could test the potential of direct air capture more seriously himself. He assembled a group of interested graduate students and postdoc researchers to do more in-depth research.

Geoff Holmes was one member of the team. He had become interested in climate change after working as a river guide on big rivers in northern British Columbia like the Tatshenshini and Stikine. Receding glaciers had become difficult to ignore, some of them having normally protruded right into the river. "It was hard to see the impact on rivers and not pick up some tools and try to do something about it," said Geoff. What tools those might be wasn't obvious though. He had graduated from university with a degree in physics but didn't have an appreciation for what role physics played in climate change, nor what role he could play. He started looking for people—atmospheric modelers, energy system analysts, technology developers, economists—who might be able to guide him. The most helpful feedback he received was to begin with the basics and learn the fundamentals of every type of energy source, starting with Vaclav Smil's book *Energy at the Crossroads.*

While educating himself, Geoff came across David Keith. What immediately stood out was that David was researching vastly different aspects of climate change, not just through an academic lens but also from the perspectives of business, industry and government. "It was pretty clear to me that I needed to study in his group, if he would have me." After several attempts, he managed to meet with David, who appreciated his earnestness and offered to help him develop an academic plan. Geoff would complete a master's degree in his new mentor's energy and environmental faculty, with a major in physics and a thesis focused on direct air capture, specifically the absorption of carbon dioxide by liquids. In September 2008, Geoff began meeting regularly in David's office with a handful of other graduate students and postdocs investigating different elements of capturing carbon dioxide from the atmosphere. Geoff sat quietly and listened. "I was completely useless at the beginning."

The work David and Klaus had done up until that point had suggested that there was nothing impossible about direct air capture, despite the reservations of many who viewed it as quixotic. Geoff still wondered how he would tackle the seemingly impossible task of separating carbon dioxide from the 2,500 times greater air mass it's contained within. He started experimenting with the use of different chemical solvents, building an air scrubber that sat atop a bench to house them. As the air was pushed through the scrubber, the carbon dioxide content was measured when it entered the unit and when it exited. David pushed Geoff and the team to use components and processes that had been proven by other industries and were readily available. Paper production, as an example, commonly includes a process where a hydroxide is used to remove impurities from pure pulp, which produces a carbonate, so-called black liquor, from which hydroxide is recovered for reuse. So Geoff began talking to people in the pulp and paper sector to find the optimal way to recover the hydroxide. He had quickly become a fully contributing member of the team, for which he credits his mentor: "David was an impeccable graduate supervisor and prof—he was really good at both."

Shortly before Geoff joined the team in 2008, the Discovery Channel had approached David about filming an episode as part of its Project Earth series and, rather unexpectedly, offered to pay for something to be built specifically for the show. Up until that point, designs for the contactor, which introduces air to the carbon dioxide–removing sorbent, simply forced upward-flowing air into the downward-flowing liquid sorbent. "The more you looked at it though, the more it was a fuck-up," said David. For the Discovery Channel, the team decided on a new approach and settled on a four-by-twenty-foot tower made with structured packing material (corrugated metal or plastic sheets commonly used in chemical reactors) that would force air circulated by a fan to take complicated paths through the tower, all while contacting a large surface area covered with carbon dioxide–absorbing liquid sodium hydroxide in the process. "We weren't thinking of it as the right approach," said David about the new prototype. "But it worked incredibly well."

By the following year, the engineers on the team had put together a process flow diagram that laid out the inner workings of their proposed direct air capture unit, along with basic cost and performance assumptions. "As you do more engineering, it becomes less ridiculous," said David. To make further advancements, however, would require brute force engineering—building prototypes, running tests, improving performance, reducing costs. "Our process had nothing super high tech and sexy that you publish articles about in *Science* and *Nature*," said David. "It's classical chemical engineering which allowed us to go big in scale quickly."

To progress any further, though, it became apparent that a company would have to be formed and money raised—something foreign to David: "You have no idea what you don't know." He reached out to more experienced businesspeople for guidance on how to write a business plan, extract patents from a university and capitalize a company. One adviser, an experienced professor who had significant academic and practical experience, became involved to help find holes in the process design. "He found tons of small holes, but in the end, he said there was nothing fundamentally stupid with what we were doing and that we should continue down the road," said David. Another had started up a successful, unrelated company and shared advice on raising money. Even Peter Eisenberger, who was now running his own direct air capture company, Global Thermostat, alongside Graciela Chichilnisky, provided advice. The outreach culminated in a meeting with all the advisers, who concurred it made sense for David to proceed and start raising money.

Naturally, David reached out to Bill Gates first. He sent him a business proposal that is now publicly available as part of a Harvard Business Case. It acknowledges the seeming absurdity of trying to capture carbon dioxide from the air we breathe compared to the challenge of capturing it from more concentrated industrial smokestacks. The proposal argued, however, that the concept was not only backed by the laws of physics and thermodynamics; it was also a scalable opportunity that provided flexibility in locating air capture plants, including in lower-cost countries like China. Even without moving operations offshore, though, the proposal suggested that with disciplined process improvement and construction management, an all-in cost of $100 per ton might be achievable, which was lower than many other carbon abatement options. David asked for $3 million; Gates said he was in for half, so long as other funders made up the difference. The Virgin Earth Challenge had helped attract some attention, but it was proving difficult to find interested investors. He found himself on a panel at an energy conference with an eminent investor in the Canadian oil patch named Murray Edwards. "Hey, I'm doing this thing—it's a little crazy, it's high risk—would you be interested?" He was, and Carbon Engineering was born in 2009.

One guiding principle for the company would be taking advantage of economies of scale. They had designed a contactor that could be scaled to an enormous size to bring the cost per ton down as much as possible. They had also selected a liquid sorbent to react with air, as opposed to a solid one as Klaus Lackner and Global Thermostat had, which would allow for greater economies of scale with respect to the post-capture processes, given the relative ease of moving large volumes of fluids around. One of the potential disadvantages of Carbon Engineering's approach was that high-temperature heat, around 900 degrees Celsius, would be needed to recover the chemical sorbent and release the carbon dioxide. As a result, they wouldn't be able to use waste heat from industrial processes. David didn't see much advantage in using waste heat, however, because it's not actually that readily available; most companies work hard at finding applications for waste heat, and if they haven't, it's because they don't want to be tied to a facility under the control of another company.

To further prove up the concept, Carbon Engineering needed some land for a larger-scale pilot project. Squamish, a postcard-worthy town north of Vancouver, offered free land on a repurposed industrial site previously contaminated with mercury; after the British Columbia government threw in some money, it became the company's new home. At the same time, the pace of development accelerated, a new CEO was hired and directors with more relevant experience were brought on board.

Meanwhile, David took on a professorship in both applied physics and public policy at Harvard University in 2011 and saw his usefulness to Carbon Engineering diminish. "I am reaching my level of incompetence in some ways . . . Some of my management weaknesses are things that I could get coaching for and fix, but I can't do that and do the Harvard job. It just isn't working," he said at the time.

David still serves on Carbon Engineering's board of directors, but he avoids any potential for conflict by eschewing carbon removal research topics at the university and making it abundantly clear when he's acting on Carbon Engineering's behalf as a director that he's not wearing his Harvard professor hat. Most of his research at Harvard has been focused on solar geoengineering. While at the University of Calgary, he had been concerned that working on solar geoengineering might be perceived as a conflict. One of the most frequently raised concerns with solar geoengineering is the potential for moral hazard: delaying emission cuts because of a diminished view of climate risk if a magic fix is deemed to exist. David believed that argument would have been bolstered if his work with energy companies included solar geoengineering. At Harvard, he felt free to pursue his interest in solar geoengineering more earnestly.

BILL GATES LABELS solar geoengineering a "break glass in case of emergency" option that might be nice to have available if the planet's climate appears to be approaching a tipping point. He provides the example of ice-like crystalline structures containing large amounts of methane that are found on the floor of the ocean. If they were to become unstable and erupt, it could lead to rapid warming, which in turn might trigger other large releases of planet-warming gases stored below the land or ocean surface. The prospect of using solar geoengineering as a short-term, low-cost fix came up in Gates's first session with David. It troubled Gates to learn that no government funds had been committed to funding research, so he, along with some former Microsoft colleagues, established a fund that David and Ken Caldeira could issue grants from. The Fund for Innovative Climate and Energy Research has given out several million in grants to numerous climate-related research projects and directly funded some research by the two scientists.

Gates's 2021 book, *How to Avoid a Climate Disaster*, begins with a concession that he is an imperfect messenger on climate change: "The world is not exactly lacking in rich men with big ideas about what other people should do, or who think technology can fix any problem. And I own big houses and fly in private planes—in fact, I took one to Paris for the climate conference—so who am I to lecture anyone on the environment?" But he argues that while techno-fixes are insufficient on their own, they're also necessary. His sizable contributions pale in comparison to long-term government spending, but they have been significant in identifying opportunities that might have been missed by governments. The co-recipient of some of his money, Ken Caldeira, says that increasing pragmatism as he grows older has left him increasingly appreciative of Gates's perspective. "Put it this way," he said. "If I had a lump of money to invest to achieve some goal, and you told me that either I could invest it myself, or the federal government could invest it for me, or Bill Gates could invest it for me, I would pick Bill Gates without a moment's hesitation."

Gates advocates a multifaceted approach to eliminating emissions—influenced by his one-of-a-kind vantage point as an investor in many germane startups that he believes will be sufficient in avoiding a climate disaster. "I've put more than $1 billion into approaches that I hope will help the world get to zero, including affordable and reliable clean energy and low-emissions cement, steel, meat, and more," writes Gates. "And I'm not aware of anyone who's investing more in direct air capture technologies." He offers up a hypothetical scenario whereby direct air capture is used to remove the equivalent of all annual global emissions.

With some additional innovation, he believes direct air capture will be feasible at $100 per ton. Multiplying that by the roughly fifty billion tons of annual emissions would result in a $5 trillion cost—roughly 6 percent of worldwide GDP or, coincidentally, about the same amount the International Monetary Fund has estimated goes towards subsidizing fossil fuels each year by governments around the world. He also acknowledges that such an approach would be extremely inefficient, given that it's generally much cheaper to just avoid emitting greenhouse gases in the first place.

But just as direct air capture won't solve the problem on its own, curtailing emissions won't either—unless we choose to ignore the harm and risk already caused by historical emissions, which will be compounded by future emissions. Carbon removal and/or solar geoengineering will be required to counteract impacts already locked in. Having thoroughly investigated both options, David has an informed perspective on the challenge. He argues that ramping up carbon removal too quickly could lead to supply chain issues that will drive costs up and that getting to a meaningful scale by midcentury, like the one envisioned by the IPCC, comes with a huge environmental footprint and social pressures. He likens the scale to the existing infrastructure that moves fossil fuels around, or nearly that of the mining sector. But he also emphasizes that solar geoengineering is not a long-term solution.

David wrote an op-ed for the *New York Times* in 2021 that laid out the circumstances. If emissions are eliminated by midcentury, it would take thousands of years for greenhouse gases to slowly be removed from the atmosphere by natural processes. What is required to diminish the threats posed by a warmer world in the meantime? The options, which were closer to science fiction when David started researching them decades previously because nobody else was, are now very real. "Yes, this is what it comes down to: carbon removal or solar geoengineering or both. At least one of them is required to cool the planet this century. There are no other options."

DAC 1

JIM MCDERMOTT WAS studying business at UCLA in 1996 when he ran out of stamps one night while mailing off job applications. Down the street from where he lived was a post office in a federal building, but it was closed. He wondered why stamps couldn't be purchased online. Two of his business school buddies were similarly intrigued, and the three set out to complete some market research during their lunch hours, interviewing passersby on Wilshire Boulevard. They put a business plan together, submitted it as part of a class they were taking specifically on writing business plans, and continued working on the idea through the summer. When they returned to school in the fall, they decided to roll the dice on starting a company to sell stamps online. They registered Stamps.com, eventually found some interested investors and received one of two licenses to sell stamps electronically on a test basis in 1999. Their company had a competitive advantage by not needing any special hardware—it was only a matter of printing off a bar code and attaching it to the package—and went on to become a multibillion-dollar company with a near-monopoly on online postage.

Jim grew up in Washington State, and his father was a physician who served in the state senate. To spend time together, Jim Jr. often accompanied Jim Sr. on work trips. That included visiting nuclear power plants that had been built by Washington Public Power Supply System (which had an acronym with an unfortunate pronunciation: "whoops"). It had been established to allow publicly owned utilities to combine resources and build four large nuclear power plants, and ultimately ended up in bankruptcy for a multitude of reasons. For Jim, tagging along with his dad, who sat on committees overseeing the plants, learning about what went into the design, construction and operation made a long-lasting impression on him. "Electricity is terrific and nuclear power is awesome," he recalled thinking. "From as early as I can remember, I just loved electricity."

Before enrolling at UCLA, Jim worked for a New York investment bank in energy finance, which typically entailed putting deals together for large coal- and natural gas-fired projects. It became clear to him that the world was on a long, slow march towards decarbonizing. "We started off with lignite coal, then we moved to anthracite coal, and then we were moving into natural gas. When I joined the bank, you could see that at some point we would move into wind and solar too." Sensing opportunity, he moved to California, which was deregulating its electricity utilities, to attend school with the plan of joining the transitioning industry. The night Jim went looking for stamps to send out résumés, he was applying for jobs to develop wind and solar projects.

With no immediate prospects, though, and graduate school debt to pay off, he was very motivated to make Stamps.com work, but before the company could begin operating, servers had to be installed to manage data. That required renting space in the first data center in southern California, alongside a plethora of others feeding the dot-com rush. As Jim walked through the maze of hardware, the implications of it all hit him: "If the whole world is going to be these banks and banks and banks of internet servers, that's a lot of electrons. And those electrons are coming from somewhere, and I happen to know that most of that's from combustion. And that means there's going to be a lot of carbon in the atmosphere as a result of this." He didn't do anything about it at the time because he was in the software business, but he kept thinking, *There is no way this works over the long term.*

In 2001, having exited from another internet company he founded and ran (a precursor to LinkedIn), and believing the internet wasn't going to solve anything, Jim decided he was going to fully devote himself to clean energy. He founded the US Renewables Group, a private equity fund focused on investing in renewable energy and clean fuel technology. As he sought out investment opportunities that reduced emissions, he realized the sole focus of all decarbonization efforts he came across was slowing down the rate of emissions growth, the rate at which new carbon dioxide was entering the atmosphere. Even in a hypothetical scenario where all emissions were eliminated, there would only be about three parts per million less carbon dioxide being put into the atmosphere each year. What about the roughly 120 parts per million already added to the atmosphere since the Industrial Revolution, which would continue impacting the climate for hundreds of years? "It's the difference between the debt and the deficit, and people weren't even getting that exercised about the deficit," said Jim. "The requirement to extract carbon dioxide directly from the atmosphere and do something with it, whether sequester it underground, or turn it into value-added products, or find uses for that carbon, was going to be a significantly larger business ultimately than just reducing the amount of emissions that we put up into the atmosphere."

Whenever Jim has pursued a new idea, he's found that there's always someone who's already been thinking about it. So, when he started digging into what to do with all the carbon dioxide already in the atmosphere, he wasn't too surprised to stumble on a 2005 research paper about capturing carbon dioxide from the air. "It is physically possible to capture CO_2 directly from air and immobilize it in geological structures," read the first sentence. While the paper suggested that trying to predict the future cost of the technology might be a fool's errand, it acknowledged that it was nevertheless worthwhile for influencing climate policy (reluctantly, it offered an estimate of $500 per ton of carbon, based on technologies available at the time). *This guy, whoever the hell he is, gets it,* thought Jim. The lead author was David Keith. When they first spoke, Jim was immediately enamored: "He was a very practical, salt of the earth guy—and smart as hell." While Jim believed David's "super-seminal thinking" would lead to something great, it wasn't a fit for his fund given the time horizon of a potential payout. "I think you're ten years too early. I recommend you find a billionaire," was Jim's parting advice.

Several years later, his partner in US Renewables Group having retired, Jim thought about starting another fund but sensed that the world wasn't yet ready for his version of direct air capture as the central pillar of sustainability. He started winding down the firm and took it upon himself to find his employees new jobs outside of the company. One of those employees ended up working for Bill Gates, who had recently formed Breakthrough Energy, an umbrella organization for several entities that were funded to accelerate technological innovation with significant sustainability potential. Shortly after settling into the new position, Jim's former employee gave him a call.

"Hey, there's this company up in Squamish called Carbon Engineering that we've got money in, and they're doing something called direct air capture. We're not sure if the market's ready for it. Would you go look at it?"

"Sure, I love that stuff," said Jim.

"It's this guy named David Keith."

"David Keith? You mean he actually did it?"

Any success Jim has had as an investor has come from betting that something was inevitable; only the timing was indeterminate. "One of the things I've always loved about David is that he understands that there's a certain inevitability if you have a deep understanding of science. Facts don't care what you think ... CO_2 is building in the atmosphere, and it has demonstrable effects that are accelerating," said Jim. He liked what he saw at Carbon Engineering and joined the board of directors and invested some of his own money. "I've got to get involved here, because this is a big part of the puzzle, which is to extract CO_2 at a very low cost. It's a problem that we need to get solving right now."

THE INNOVATOR'S DILEMMA, written by Clayton Christensen, is a book that explores the concept of disruptive technologies. In it, the author argues that there are two principal reasons why new technologies cause great firms to fail. First, the gains from innovation for a new technology typically produce minimal gains for companies at first, then exponential gains, and then minimal gains again as the technology matures. Second, incumbent firms tend to continue embracing mature technologies, even after innovation gains have become minimal, because there are high expectations from shareholders to meet annual sales targets. Existing technologies are viewed as a safer bet to meet those expectations. As a result, new entrants that don't face the same expectations often end up with an advantage because they can focus on disruptive technologies.

The book is one of Jim's favorites. He likes to point to a friend who failed at transforming an incumbent to illustrate the theory. David Crane was hired by NRG Energy in 2003 as it emerged from bankruptcy after overextending itself. He grew the electricity generation company substantially and made many investments in solar, wind and nuclear. In 2014, Crane wrote a Jerry Maguire–type letter to shareholders in which he laid out his plan to transform the company from generating most of its income from coal- and natural gas-fired power plants to renewables:

> The day is coming, when our children sit us down in our dot-age, look us straight in the eye, with an acute sense of betrayal and disappointment in theirs, and whisper to us, 'You knew... and you didn't do anything about it. Why?' And for a long time, our string of excuses has run something like this: 'We didn't have the technology... it would have been ruinously expensive... the government didn't make us do it...' But now we have the technology—actually, the suite of technologies—and they are safe, reliable and affordable.

Crane was fired the following year. NRG stock was performing poorly at the time, but ironically it was the firm's coal-fired assets that were struggling. It didn't matter: "There's no market on Wall Street for internal transformation," Crane said in an interview after being let go.

From Jim's perspective, it's the same for oil and gas companies. He believes there was no incentive to change course because there was no incentive to reinvent the business model itself. Costly research and development programs that might take decades to pay off were difficult to justify for oil and gas companies competing on cost in commodity businesses. The most impactful oil and gas innovation in recent history, hydraulic fracturing, was commercialized with the help of U.S. government-funded research and development following the energy crisis of the 1970s. The Department of Energy recognized that unconventional natural gas resources held significant potential for overcoming declining natural gas production and devoted resources to accumulating knowledge and data and testing new drilling techniques. An enterprising geologist named George Mitchell, whose company capitalized on the work done by the Department of Energy, is generally credited with first commercializing the technology that resurrected the U.S. oil and gas industry. But those who worked with him recognized the footsteps they were following in. "[The Department of Energy] did a hell of a lot of work, and I can't give them enough credit for that. DOE started it, and other people took the ball and ran with it. You cannot diminish DOE's involvement," said Dan Steward, former vice president of Mitchell Energy.

There were several reasons behind the creation of the Department of Energy in 1977; researching options to address climate change was one of them, even if the term wasn't yet in use. "The basic problem is that there is no constituency for an energy program," said the first secretary of energy in the U.S., James Schlesinger. "There are many constituencies opposed. But the basic constituency for the program is the future." Shortly after the creation of the Department of Energy, Schlesinger launched the Carbon Dioxide Effects Research and Assessment Program. It was a comprehensive plan to address increasing carbon dioxide levels. "The doubling of atmospheric CO_2 is likely to occur during the middle of the next century if use of fossil fuel continues to grow," stated a 1980 report from the program. "Present-day concepts of environmental control technology of CO_2 do not appear feasible from economic and energy requirement standpoints." Four decades later, the Department of Energy is starting to put money into direct air capture. But one CEO is not waiting to see the results. She's already begun transitioning her company.

There are two things Vicki Hollub is passionate about: football and the Permian Basin. Growing up in Alabama, she worshipped the Alabama Crimson Tide football team and especially their revered coach, Bear Bryant. She studied whatever she could from afar—his mindset, his plays, his leadership tips—and even ventured onto the field with a French horn in the marching band. Music wasn't going to lead her to a career though, so she went to work on rigs in Mississippi after graduating with an engineering degree and returned to Tuscaloosa for home games. Shortly thereafter she began working for Occidental Petroleum, where she has remained for the last four decades, becoming CEO in 2016. She credits her Crimson Tide passion for generating a commonality with male colleagues and helping her to thrive in a very male-dominated business.

The Permian Basin is the largest petroleum-producing basin in the U.S., and one of the oldest, having produced oil for over a century. "I think the Permian is going to be the last basin standing... and Occidental will be the last company standing in the Permian," Hollub says in a promotional video. For decades, Occidental has been injecting carbon dioxide into depleting oil reservoirs to stimulate more production of oil in the Permian, which stretches across west Texas and southeastern New Mexico. There are thousands of miles of pipelines transporting carbon dioxide from underground domes housing carbon dioxide emitted by volcanoes or from processing plants where it is separated from natural gas to over 6,000 enhanced oil-recovery injection wells. Hollub's assessment is echoed by Jim: "The Permian Basin is an amazing place in terms of infrastructure for CO_2. Occidental has billions of dollars in infrastructure and the most well-developed CO_2 operations. And they're not a supermajor, so they have to be more innovative."

Carbon Engineering's first commercial project, DAC 1, owned and operated by Occidental, will sit on about a hundred acres in the Permian Basin. It is expected to cost between $800 million and $1 billion and capture over half a million metric tons of carbon dioxide per year—a significant jump from the world's existing capacity of several thousand tons per year—once it's up and running in 2024.[4] The carbon dioxide captured by DAC 1 will remain underground for a very long time, but it will also release

---

4 The cost estimate includes pre-investing in making the facility scalable to one million metric tons per year. The estimated levelized cost of capture including storage for DAC 1 is estimated to be $300 to $425 per ton, according to Occidental's March 2022 investor presentation.

oil that remains in the reservoirs by infiltrating rocks and acting like soap on a greasy pan to release more stubborn oil. How much additional carbon dioxide will be added to the atmosphere from the released oil will vary depending on reservoir-specific characteristics.

Enhanced oil recovery is one of many uses for carbon dioxide. Some other uses of the slightly pungent, acidic-tasting gas—manufacturing fertilizer, carbonating beverages, decaffeinating coffee—typically rely on sourcing supply from corn-to-ethanol plants, where carbon dioxide is produced as a by-product. Additional applications, as an ingredient in building materials and apparel for example, are under development with the intent of finding a purpose other than sending it below ground. Like Global Thermostat, Carbon Engineering has another product in mind, a synthetic fuel not much different from gasoline, produced by mixing carbon dioxide with hydrogen. The hydrogen feedstock can be sourced by separating water into its constituent elements using electricity. So long as the whole process is powered by clean energy, the product is a clean fuel that can be poured directly into existing gas tanks. That's an advantage over fuel cells, which create electricity as a by-product of combining hydrogen and oxygen and require new infrastructure, including the replacement of gas tanks with storage tanks that can handle highly explosive compressed hydrogen. From Carbon Engineering's perspective, a synthetic crude could be particularly appealing for applications that are difficult to electrify, like aviation and oceangoing shipping. "A world-changing product that doesn't require the world to change," said the company's CEO, Daniel Friedmann.

Before Carbon Engineering can fully entertain different sales channels, it needs to bring the cost of direct air capture down, which has required finding a buyer. While some critics have viewed the use of captured carbon dioxide to lengthen the lives of existing oil reservoirs as an unwarranted prolongation of the fossil fuel era, the company viewed it as an important interim step on the path towards commercializing the technology (increasing the longevity of existing oil reservoirs using enhanced oil recovery is also more environmentally benign than finding and developing new ones). Beyond the enhanced oil recovery market, there were a couple of other key factors that led Carbon Engineering to Occidental. In 2018, the U.S. passed, with bipartisan support, a rule referred to as Section 45Q that expanded a tax credit made available in 2008 for carbon capture; each ton of carbon dioxide captured and sequestered would receive a credit of $50 per metric ton, or $35 if applied to enhanced oil recovery projects.

The most impactful incentive, though, came from a rule enacted by Arnold Schwarzenegger when he was governor of California: a low-carbon fuel standard introduced in 2007. It remains in place with a target of reducing the carbon intensity of transportation fuels by 20 percent by 2030. Fuel providers who can't meet the carbon intensity threshold can purchase credits—which have traded for nearly $200 per ton since DAC 1 was announced—from less carbon-intensive providers and direct air capture plants. Credits for low-carbon fuels can be sold to anyone, so long as the fuels make their way into California. Direct air capture plants, however, can be located anywhere in the world following a 2019 amendment. Jim gives the former governor a lot of credit: "In all honestly, when Schwarzenegger was elected, I thought he was the biggest meathead ever. But when you look at the substance of the 2006 [Global Warming Solutions] Act that he created and passed, it is the underpinning for the California Low-Carbon Fuel Standard, which is the single biggest piece of climate legislation ever passed. Schwarzenegger made that happen. For all his body-building craziness, you simply can't argue with the substance."

Hollub has previously been coy about transitioning Occidental to a carbon management company, possibly because of the innovator's dilemma. That was apparent when she made another move to transform the company. In 2019, Occidental won a protracted battle to take over another large oil producer, Anadarko Petroleum. The biggest prize was an even greater presence in the Permian Basin; only a brief remark was included about carbon dioxide, buried towards the end of a press release announcing the proposed takeover. It tepidly referred to how the deal "leverages Occidental's existing initiatives to utilize its CO_2 enhanced oil recovery expertise and infrastructure for economic and social benefit by applying its low-carbon strategy to Anadarko's asset base." The deal came at a hefty price: a roughly 60 percent premium to Anadarko's share price (Occidental also outbid another suitor, Chevron, and had to pay a $1 billion breakup fee to null that deal). Furthermore, to get enough cash to make the deal enticing, Hollub had made a visit to Warren Buffett. It took one meeting with him, on a Sunday morning at his nondescript Omaha headquarters, to receive a commitment for $10 billion in financing. Afterward, a critical Occidental shareholder, billionaire Carl Icahn, lambasted Hollub, seemingly aghast at the 8 percent preferred share rate Buffett received: "She ran to Omaha to make this deal with Buffett... Usually Buffett goes with Charlie Munger [Buffett's right-hand man] before he does a deal. He didn't even bother on this one. He just did the deal in ninety minutes... The whole thing is a travesty."

Investors were also unhappy with the Anadarko deal, and Occidental's share price hit a ten-year low, dropping from over $80 a year earlier to less than half that. Then the COVID-19 pandemic arrived, and it didn't take long for oil prices to tank as the world began shutting down in various ways. Hollub was still trying to sell assets to help pay down the heavy debt burden taken on to finance the Anadarko deal when Occidental's share price hit $10 in March 2020. "It may have seemed like five years to you," Hollub said in response to a sympathetic question during an interview with NBC about what 2020 felt like, "but it was more like ten or fifteen to me. It was a really tough year."

There's one Bear Bryant quote that Hollub likes to use: "It's not the will to win, it's the will to practice and prepare for the game." For her, approaching every day with that mentality leads to being prepared for crisis, or opportunity. She is no longer coy about transitioning Occidental into a carbon management business. She expects Occidental's Scope 1 and 2 emissions (those emissions not related to end-product use of oil and gas) to be net negative by 2040 and, much more ambitiously compared to its peer group, net negative for Scope 3 emissions (which include end-product use) by 2050. And she wants to help other industries—maritime shipping, aviation, cement, steel—become net negative by sequestering carbon dioxide in the Permian Basin. The company, which had previously announced a plan to build twelve large direct air capture plants in the Permian Basin, is now planning to build seventy DAC plants throughout the world by 2035 (contingent on supportive public policy, customer demand, suitable storage, mature supply chain, etc.). "It's probably going to be a $3–$5 trillion industry if you look at how much carbon capture is going to be needed around the world," Hollub said. "We think, ultimately, it's going to generate as much earnings and cash flow as our oil business does today. We believe it's a long lasting busi ness." Warren Buffett agrees. His holding company, Berkshire Hathaway, purchased another large stake in the company, about $7 billion in common shares, after Buffett read a transcript of a conference call Hollub held with investors in February 2022. "I read every word, and said this is exactly what I would be doing. She's [Hollub] running the company the right way," he told CNBC.

"It's game on," says Jim. "Just when you thought the Americans had their heads all the way up their asses and appear to be nothing but an impediment to climate solutions, a new industry is born via new technology and creative destruction." And he doesn't think the capacity to sequester carbon dioxide will be a limiting factor, pointing to subsurface reservoirs in Texas, Louisiana and the Gulf of Mexico that may alone have sufficient capacity to sequester enough carbon dioxide to return the atmosphere to 280 parts per million of carbon dioxide. "Texas is going to win," argues Jim. "CO_2 extracted in Texas can be sold anywhere; it has the regulatory advantage [of U.S. Section 45Q tax credits in addition to the California Low-Carbon Fuel Standard credits available anywhere]; it has all of the infrastructure already set up for moving CO_2 around, which can be redeployed; and it has access to the world's largest capital markets. Texas might be the lowest-cost hub for direct air capture."

It leaves Jim sanguine about the situation. "When you get into it with people who are actually in the oil and gas business, they're like, 'We can solve this.' It's not a technical issue per se, it's not that more science is required, it's about money and political will." And he believes the U.S. may have already closed the gap for a good portion of the carbon that could be removed in the next five to ten years. As demand for carbon renewal ramps up significantly from other markets, both from measures already in place like increasing carbon prices and from new measures still to be introduced, he believes the constraining factor will be the technology providers' bandwidth.

David Keith sees other obstacles for technology providers like Carbon Engineering: capital constraints, supply chain logistics, environmental footprint, public acceptance. Looking at just the electricity that could be required to power DAC plants alone is staggering. By Jim's quick math, if there are 10,000 DAC plants removing a total of ten billion tons per year of carbon dioxide and each one requires the equivalent of about sixty to one hundred megawatts of electricity, then that's equivalent to up to a million megawatts of capacity, roughly the size of the entire generating capacity in the U.S. today. That excludes other growing uses of electricity as additional components of everyday life get electrified. To meet that demand, Jim believes that wind and solar will have to continue growing like mad, and a renaissance in nuclear power is forthcoming. The land footprint for nuclear is a small fraction of what it is for wind and solar, and further innovation will continue to bring costs down and safety performance up. "How many people have actually died from nuclear versus how many people might die from heat?" he asks. "The answer is clear: excessive atmospheric CO_2 is way more dangerous than managing nuclear." Building more generating capacity, in his opinion, will help economies grow and enhance well-being. Conversely, dampening the rate at which direct air capture is deployed to ease its adoption will lead to exponentially greater problems down the road.

"I made the mistake a couple of times of arguing with David about technical things—so I don't," said Jim. But when it comes to bigger-picture matters like the rate of removing carbon from the atmosphere, he strongly believes that haste is appropriate. "My view is we are in such a corner environmentally, from a carbon perspective, that we need to go very fast. And I'm not sure what downside there is to going fast and finding out that some of the things we did in going fast have negative outcomes—versus the certainty that, if we go slow, it's going to be horrible."

What if fast is still not enough? We might be thankful that David spends as much time thinking about contingency plans as he does, which in this case might be solar geoengineering. "Here's what I think is really going to happen," said Jim. "DAC is going to grow like mad. But it is not going to grow fast enough. And sometime—I'm guessing around 2035 to 2040—we're going to realize that we need to use the Band-Aid." At that point, he believes many components of the green economy will be ticking along and there won't be an aversion to using a quick and cheap solar geoengineering fix to buy time. "I spend a lot of my time in the future, and David seems to be one of the few people I've met in my life who seems to be ahead of me in the future, consistently. I find that quite comforting."

ORCA

ETH ZURICH, ALSO KNOWN AS the Swiss Federal Institute of Technology in Zurich, is one of the most highly regarded universities in the world. It's a top choice for aspiring engineers and scientists. Yet tuition is inexpensive at about $800 per semester and admission doesn't even require a high school education; passing an entrance exam can suffice (although not even Albert Einstein was able to in 1895). For Christoph Gebald and Jan Wurzbacher, it just seemed like a nice place to study. The two met on the first day of classes in 2003 and not only bonded over their mutual struggles with Swiss German but also discovered a shared fondness for climbing, skiing and entrepreneurship. That same day they decided they would start a company together, formalizing the commitment with a high five.

Six years later, after completing Master of Engineering degrees and climbing and skiing together on glaciers retreating at a pace that rattled them, the pair finally landed on what their company would do: capture carbon dioxide from the air. "We didn't want to develop an app or create a platform," said Jan. "We just wanted to build a tangible machine that solves the problem." A professor named Aldo Steinfeld had inspired them with his research. Steinfeld had a vision of using solar energy to produce renewable fuels using Klaus Lackner's concept to capture carbon dioxide from the air and hydrogen from water. Jan and Christoph wondered if they might be better off focusing on just direct air capture. That still left them scratching their heads: How would they overcome the seemingly insurmountable problem of the diluteness of carbon dioxide? "That is the art of direct air capture," said Jan. "Klaus Lackner was the first to come up with that concept on a larger scale."

As they experimented in Professor Steinfeld's lab with the same scrubbers found in submarines, they also completed an entrepreneurship course. At its outset, students were asked to propose a business idea, then everyone would vote and a handful of ideas would be transformed into business plans with help from the rest of the class. The pair's direct air capture idea garnered the most votes, and other students were assigned to help with the plan. "We thought about markets, about intellectual property, about financing, about technology—well, not so much about technology actually because that was rather our part," said Jan.

One of the students assigned to them said he knew of a greenhouse outside Zurich that needed carbon dioxide. When Jan and Christoph ventured to the nearby town of Hinwil to investigate, they met a family who had started up a greenhouse operation and were indeed looking for low-cost carbon dioxide to enhance the yield from their vegetables. Conveniently, nearly next door to the greenhouses was the town's waste incineration plant, an ideal source of heat. The pair set to work designing a prototype. Professor Steinfeld gave them each a 50 percent PhD position that allowed them to continue using the university's labs and other resources, and a stipend to split. Meanwhile, the other 50 percent of their time could be devoted to their new business—they named it Climeworks in homage to their favorite bootfitter, a small shop in Chamonix named Footworks.

The first prototype the pair built in 2010 was ungainly: an aluminum bucket containing a couple of hoses pushing air over filters coated with amines to absorb the carbon dioxide. It barely worked, taking a full day to capture about half a gram of carbon dioxide. It was, however, enough to earn about $130,000 in prize money from a Swiss fund established to kickstart university spinoffs. A month later, David Keith fortuitously showed up in the lab after becoming interested in some research Professor Steinfeld was doing. It didn't lead to anything; instead, Jan and Christoph took him out for a beer to talk about their new venture. David, who had founded Carbon Engineering the year prior, was impressed by the work they had done and encouraged them to continue. "It was a really cool time. We were like, 'Yeah, we've got a company—we're going to solve it,'" said Christoph. Most responses to their pursuit have been positive. "When I talk to a random person on the street asking me what I do, and I say I run a business that captures CO_2 from the air, nine out of ten people say, 'Hey, that's great!'" said Jan. "Because everyone knows that CO_2 is a problem, but they are confronted with various complex schemes for tackling it." Some who didn't show support questioned whether the approach would create too much of a moral hazard, enticing people to continue emitting if they knew the carbon dioxide could later be later cleaned up by somebody else.

Of more immediate concern, as Jan and Christoph sought to ramp up their business, was a 2011 report from the American Physical Society that suggested that direct air capture, at an estimated cost of $600 per metric ton, was too expensive to justify pursuing anytime soon (the same report that created headwinds for the other direct air capture startups). It was read by prospective investors in Climeworks, who were being asked to invest in an unknown technology—being developed by a pair of unproven university graduates—and it was saying that the technology might not be cost-competitive with other options until later in the century. But one thing working for the pair was the Virgin Earth Challenge, which was helping to raise awareness, showing that scientists with much more credible track records were also getting behind the technology. Investors eventually agreed to put $2 million into Climeworks, with a condition: a prototype capable of capturing one kilogram of carbon dioxide per day was to be built by the end of the year.

After a series of trials and errors, they had built a contraption the size of a refrigerator by mid-December. But when they tested it, the reading fell short of expectations: it was showing only about 200 grams over the course of a day. They continued tinkering, but nothing was helping. Just before Christmas, as Jan despairingly watched the unit run, he picked up on an unexpected hissing sound. Upon further inspection, he discovered that one of the hoses carrying carbon dioxide was not properly fastened. After Jan fixed the problem, the sensor showed that they were in fact capturing several kilograms of carbon dioxide. They had met the financing condition.

To build a commercial facility, they moved into a larger workshop and designed a semi-automated manufacturing line for the eighteen modules that would collect carbon dioxide for the greenhouses. Meanwhile they opted to start small and gradually increase the size of their projects. In the process of starting up, they frequently compared notes with David Keith; Jan visited Calgary and Squamish and they hosted David in Zurich. He told them why he thought a larger scale would win out, at least with respect to economics. One advantage of the approach Jan and Christoph envisioned was that it would be more conducive to a startup: whereas Carbon Engineering would have to wait until 2024 to demonstrate the cost and performance of a full-scale plant, Climeworks could do it more incrementally. They would gradually work their way down the cost curve with each iteration of their more modular design, raising money and earning revenue along the way. "To be clear, as a business plan, theirs is way better," said David.

Still, they didn't complete the Hinwil facility until 2017, nine years after their initial conversation with the family owning the greenhouse. The assembled capture units resemble stacks of large air-conditioning units, which push air through the contactors containing a solid sorbent. When the contactors are heated with waste heat from the incineration plant, captured carbon dioxide is released and piped across a field into large greenhouses. The extra carbon dioxide accelerates the photosynthesizing of vegetables—eggplants particularly thrive on concentrations reaching 1,000 parts per million—boosting yields by about 20 percent (the flipside is that scientists have found that food grown at elevated carbon dioxide levels is less nutritious). The overall capture costs were about $600 per metric ton, and it only captured 900 metric tons of carbon dioxide per year, equivalent to the annual emissions of roughly 200 cars, but they had proved their design could work. Or at least they had proved the capture component could work. Carbon dioxide cannot be permanently stored in eggplants, cucumbers and tomatoes; the carbon dioxide in food is dissolved in the blood, carried to the lungs by circulation and breathed out. To find a permanent home, Jan and Christoph would need to look elsewhere.

Around the same time, a United Nations Climate Change Conference, COP22, was being held in Marrakesh, Morocco. Climeworks's progress had brought them more credibility, so Christoph received invitations to get-togethers with researchers, activists and policymakers. One invite he received was to a lavish party thrown by Laurene Powell Jobs, the widow of Apple founder Steve Jobs and a climate philanthropist. As Christoph made the rounds, feeling a little out of place, he came across an outgoing man who was keen to hear his story. The man's name was Ólafur Ragnar Grímsson, and he had just retired as the president of Iceland. "That's fantastic!" said Grímsson upon hearing of Climeworks. "I can store CO_2 underground in my country. But we've been lacking the technology to capture it."

WHEN ÓLAFUR RAGNAR GRÍMSSON delivered his first annual address to the people of Iceland as their president in 1997, he emphasized the importance of dealing with climate change. Specifically, he referred to the work of Wally Broecker at Columbia University, who he had come to trust as a wealth of knowledge despite not having met him. Seven years later, Grímsson was attending a climate change conference at Columbia where Broecker was presenting findings on the impact of climate change on polar environments. When the two spoke, Grímsson was surprised to learn that Broecker, along with his colleague Klaus Lackner, was interested in Iceland's potential for sequestering carbon dioxide in basaltic rock. To further the discussion, Grímsson invited the Columbia professors to Iceland in 2006 and asked Broecker to deliver the first Presidential Lecture, a lecture series he had recently initiated.

The problem with using conventional oil reservoirs, like those found in the Permian Basin, to store captured carbon dioxide is the lack of certainty it will remain there. After being injected, it is contained below an impermeable cap rock but remains a gas. Over time, the carbon dioxide dissolves into pore waters, causing it to lose its buoyancy, which eliminates the risk it will escape to the surface. But the process can take hundreds or thousands of years, so reservoirs must be monitored in the interim. Klaus had written a paper in 1995 along with colleagues from Los Alamos—inspired by his thought experiment with self-replicating auxons—that examined the possibility of safe and permanent storage of carbon dioxide by combining it with abundant raw materials to form carbonates. Just as his direct air capture concept would mimic nature as an artificial tree, accelerated mineralization could also mimic nature on a vastly shorter time scale.

Typically, carbon takes a long time to make its way through the slow geological cycle—somewhere between 100 million and

200 million years.[5] Carbon dioxide in the air dissolves in moisture to form a weak acid, carbonic acid (known in other contexts under many different names, including carbonated, soda, fizzy or sparkling water) that falls to the surface as rain. The acid dissolves rocks, a process referred to as weathering, which releases ions, including calcium ions, that are transported to the oceans by rivers. With the help of coccolithophores, corals, pteropods and other shell-bearing organisms, those calcium ions combine with bicarbonate ions to form the calcium carbonate found in shells. When the shell-bearing organisms die and sink to the bottom of the ocean, layers of shells eventually cement over time to form limestone rock. Some of the carbon in that rock is returned to the atmosphere by volcanic eruptions, which release tens or hundreds of millions of tons of carbon dioxide every year—a small fraction of the amount released by the combustion of fossil fuels.

To mimic the natural mineralization process on a shorter time scale would require rocks that aren't already carbonated (as is limestone) and that are rich in minerals like calcium and magnesium that bind with carbon dioxide. One candidate rock Klaus suggested was basalt, a type of volcanic rock with a high concentration of minerals. If carbon dioxide is pumped directly into basalt, theorized Klaus, the carbonation process might be accelerated from millions of years to just a few years. Basalt is found throughout the world but mostly on ocean floors; only about 5 percent of land area is basaltic. Iceland, however, is basically an entire island of basalt.

When Broecker was invited by President Grímsson to speak to Icelandic scientists in 2006, he spoke about Klaus's hypothesis, which had been tested by a colleague at Columbia, a young scientist from Switzerland named Juerg Matter. Conveniently,

---

5 Faster carbon cycles include annual cycles involving photosynthesis, decadal cycles involving vegetative growth and decomposition, and centurial cycles involving the ocean and the atmosphere.

Matter's office at Columbia's Lamont-Doherty Earth Observatory, located upstream on the Hudson River from Columbia's Manhattan campus, was situated atop basalt cliffs called the Palisades. They were formed over 200 million years ago when magma was thrust upward into sandstone and exposed over time as the softer sandstone eroded. Matter had injected carbonic acid in a well drilled into the basaltic rock and then returned a week later to pump water out of the well. He was encouraged to find that calcium had indeed been released from the rock after reacting with the carbonic acid to form calcium carbonate—evidence of accelerated mineralization.

After watching the presentation keenly from the front row, Grímsson invited his guests to lunch a few days later at his presidential residence, Bessastaðir, a collection of white houses sitting on a windswept, basalt-exposed peninsula across a bay from Reykjavík that has hosted Icelandic rulers since the Middle Ages. Adorning the residence's walls were paintings depicting Icelandic scenes, and Grímsson drew attention to one particular painting: an early 20th-century depiction of Reykjavík choked in smoke. At that time, most of Iceland's energy was generated from coal imported from Britain, and Grímsson proudly explained how the country was now largely powered by geothermal and hydroelectric resources. His guests left knowing that any efforts to sequester carbon dioxide in Icelandic basalt would have the president's firm personal support.

At the time, Iceland was caught up in a spending spree fueled by aggressive borrowing by banks that were privatized just before the global financial crisis. Some of the ensuing expenditures were whimsical, like the billionaire owner of one bank purchasing a Premier League football club. Others were more strategic, like Iceland's main supplier of electricity, heat and water, Reykjavík Energy, investing in a 303-megawatt geothermal power plant called the Hellisheiði Power Station, which was built largely to power large aluminum smelters. While 99.5 percent of the electricity-generating steam brought to the surface at Hellisheiði is water, about 60 percent of the remaining gases is carbon dioxide. The Columbia academics were able to negotiate an agreement with Reykjavík Energy to supply some of that carbon dioxide for the experiment. And, with input from Klaus, a way to separate the carbon dioxide from the steam was devised so that it could be returned below the surface to test the basalt's hospitality. It was hoped the initiative, dubbed CarbFix, would demonstrate it could be done economically.

CarbFix almost came to a premature end when Iceland's spending spree was abruptly halted by the global financial crisis. President Grímsson remained in power though—remarkably, he stayed in office for twenty years, until 2016, despite Iceland experiencing the largest banking collapse of any country relative to the size of its economy—and he continued supporting the project. Before large-scale testing finally began in 2014, initial computer modeling had suggested it might take up to ten years for the injected carbonic acid to react with the basalt and for the carbon dioxide to mineralize. The test began by injecting CO_2-charged water a half mile below ground and then waiting to see what would turn up in a nearby monitoring well. It only took a few days for some of the injected water to show up, having found cracks to travel through. Test results were increasingly acidic, which was a concern as that suggested the hoped-for reaction was not occurring: minerals were not reacting with the carbon dioxide. A much larger volume of water, however, was still making its way through the rock by way of tiny pores and fissures. When that water, which contained tracers to confirm it was indeed the injection water, arrived at the monitoring well, the test results returned to being less acidic; the carbon dioxide was reacting with the minerals in the basalt and turning into carbonate.

It took less than a year for the basalt to absorb over 95 percent of the injected carbon dioxide, which far exceeded the team's best expectations. Furthermore, the cost of drilling, injecting and monitoring was estimated at only about $5 per ton. The storage capacity of Iceland's basaltic rocks is estimated to be over 2.5 trillion tons, about fifty times global annual emissions. The proximity to seawater in Iceland, which can be used instead of fresh water to form carbonic acid, is another attractive feature. Reykjavík Energy was so impressed by the results that sequestering carbon dioxide has become standard operating procedure at Hellisheiði. Climeworks was also impressed and decided that it would be an attractive location for their next project.

SHORTLY AFTER 6 PM on September 9, 2021, the world's largest commercial direct air capture facility began scrubbing the sky. Looking on from the Hellisheiði geothermal plant were about 200 spectators—journalists, investors, researchers—to commemorate the event. Jan and Christoph, now thirty-eight years old, took the stage in front of the crowd. They continue to jointly manage the company, with Jan taking on more responsibility for the company's operations and finances and Christoph more of the sales and marketing duties. "This year could turn into a turning point in how climate change is perceived," said Jan. "Thirty years down the road, this can be one of the largest industries on the planet," said Christoph. Iceland's former president, Grímsson, was also on hand. "Future historians will write of the success of this project," he said.

The Climeworks facility, named Orca after the Icelandic word for having the energy to do something (spelled *orka*), was built with the capacity to capture 4,000 metric tons per year. There was a learning curve to adapt to the harsh Icelandic climate, which the facility was designed to blend in with, including natural colors and wood cladding. Its completion was an important step along the company's intended path of commercializing the technology on a grand scale. And importantly, there is now a market for it. As Orca's fans began spinning, roughly two-thirds of the carbon dioxide the facility is expected to remove and sequester underground over its lifetime had already been sold to corporations and private individuals. "Klaus Lackner's first ventures were probably just a little too early," said Jan. "He knew that we needed them, but the market was not there yet."

Climeworks has set its sights on contributing to a midcentury target put forth by some climate scientists to capture roughly ten billion tons of carbon dioxide per year globally. While direct air capture will not be the sole carbon removal option for meeting ambitious carbon removal targets, Jan expects it will play a large role as other options may struggle with scaling up or land use issues. And he hopes other new ventures will be successful if the midcentury target is to be met: "It's not one company, it's not two companies, it's likely not ten companies that can do it." And he expects to continue looking to Klaus for inspiration. "He wrote a bunch of very nice papers analyzing the fundamental physics of direct air capture which are quite important," said Jan. "As we move into a phase of optimization and running down the cost curve, then all these fundamental aspects are very important, and I think he has done very good work there."

If one assumes that a company like Climeworks can achieve a growth rate similar to other industries, such as solar panels or wind turbines, that suggests capacity could grow by a factor of ten over a decade, according to Jan. If one also assumes Climeworks needs to reach one billion tons per year of capacity by midcentury, that leads to sizable interim targets: "We need to be at a hundred million tons by 2040 and ten million tons by 2030." To ramp up, the company is busy developing the next project, which will be ten times larger than Orca and potentially in service by 2024. At the same time, they plan to branch out to other locations. Like Oman, for example, which has peridotite formations consisting mostly of silicate minerals that the company believes could mineralize trillions of tons of carbon dioxide. The country's vast, sunbaked deserts are also expected to house an increasing number of solar panels.

Climeworks's carbon removal service, which they began offering online when they became overwhelmed by requests for it, has now become an important source of revenue. In addition to corporate buyers, over 10,000 private individuals have subscribed online for carbon removal at a price of roughly $1,200 per metric ton. Some have sought to offset their entire annual footprint, at an estimated cost of about $15,000 for the average North American. Klaus bought a few thousand dollars' worth so he can drive carbon-free, briefly. Bill Gates, the largest shareholder of Climeworks's competitor Carbon Engineering, subscribed. To realize the ambitious growth trajectory the company is striving for will require many more voluntary corporate and personal purchases of carbon removal. "High-quality solutions will be needed and they cost a bit more now, but we need to develop them because they have much more scaling potential than other solutions," said Jan. "The voluntary markets could lead us to millions of tons—maybe ten million, maybe fifty million." From there, he believes regulations like carbon taxes will be required to make the leap to billions of tons.

SOLAR-POWERED PATHWAY

DAVID KEITH HAS PUBLISHED papers on the futility of forecasting the future cost of technologies, so his response to the question of what direct air capture might eventually cost is not unexpected. "It's not an answerable question," he confirmed. "We don't have a method to estimate the costs of technology in the future. And it doesn't matter for any decision we're making now because decisions get made sequentially."

The track record of forecasting the costs of environmental technologies has indeed been subpar. Most often, the costs turn out to be significantly lower than expected because of unforeseen impacts from innovation and new technologies. One of the most widely cited studies looking at cost estimates for proposed environmental regulations showed that out of twenty-one proposed regulations, costs were overestimated in thirteen cases and underestimated in only three. Acid rain compliance costs in the U.S., which were projected to total $2–$5 billion a year, ended up costing less than $1 billion per year. Estimates for controlling volatile organic compounds of $15 billion per year were later revised to less than $1 billion. Estimates for controlling air pollution from coke ovens in the steel industry dropped from $4 billion to $400 million. And sulfur dioxide regulations imposed on electric utilities that were expected to cost $4–$5 billion led to technological improvements and switching to low-sulfur coal, which resulted in cost savings of hundreds of millions of dollars.

One of the most remarkable examples of constant innovation driving down the cost of a new technology is solar energy. In 2008, David published a paper with two colleagues that advised governments to be cautious about subsidizing large buildouts of solar panels, also known as photovoltaic solar energy, despite ongoing cost reductions and performance improvements. Based on their surveying of experts, the authors concluded that photovoltaic technology "may have difficulty becoming economically competitive with other options for large-scale, low-carbon bulk electricity in the next forty years." It was a lesson in the impossibility of forecasting the impact of innovation. "I was wrong," said David. "I was worried that deployment incentives would simply lock in the current technology and do little to drive the breakthroughs that were needed to get solar cheap enough to compete."

When David was interviewing solar experts in 2007, Greg Nemet was completing a PhD in energy and resources at University of California, Berkeley. He attended a workshop at Carnegie Mellon, where David was a professor, and presented some of his doctoral thesis work. It was the first time he benefited from David's feedback: "His comments are always sharp, sometimes intensely so, and they have focused my own thinking." A few years later, Greg, now a professor at University of Wisconsin-Madison, completed a survey, along with colleagues, of sixty-five solar experts to test whether a consensus existed on the future cost of solar panels. The experts, regarded as the most knowledgeable in the field, were specifically asked to predict where solar prices would be in 2030, which seemed to Greg like a reasonable time frame for when solar might become competitive with other forms of large-scale electricity generation. Nearly all the cost estimates were beaten a decade earlier than expected. The International Energy Agency announced in 2020 that in sunny regions, solar had become the cheapest electricity in history and was roughly 20-50 percent cheaper than the agency had forecast just the previous year. Overall, between the time solar energy was first used to power a satellite in 1957 until now, the cost of solar photovoltaic panels has dropped by an astounding factor of over 10,000.

Greg's interest in solar began while working for a think-tank in Silicon Valley during the dot-com days of the late 1990s. He worked on a study that compared industries that made notable breakthroughs, such as information technology, biotechnology and consumer products, with other industries that appeared more stagnant, like energy. Part of what they were looking for was how industries compared when it came to factors that might influence innovation—money spent on research and development, number of scientists and engineers working on innovation, number of startups, number of patents applied for. What stood out to Greg was that the energy sector measured lower on every metric, and by orders of magnitude. Seeing that, and the potential for greater innovation in the energy sector to address its many challenges, he decided to pursue graduate studies to better understand how energy innovation could be better incentivized.

Greg decided to focus his work on solar energy. "It was a clean technology and had some elegant physics behind it," he said. "But people who worked in the energy system wouldn't take it seriously. The energy system is just massive and it's slow to change. And the idea that some really expensive, cool technology could actually make an impact was just dismissed." That was despite the dramatic gains already made. To better understand what was driving costs down and performance up, Greg first began working with econometric models that tried to isolate contributing factors such as research spending and efficiency improvements. He was presenting his findings one day when he came to a slide showing how the heavy lifting of investment in solar had moved around the world—from the U.S. to Japan to Germany to China. The simple fact that major investment shifted as solar matured suddenly struck him as something data alone couldn't explain.

Greg went to a high school that was, in his words, very old-school. Roxbury Latin School is a small institution in Boston that was founded in 1645 and instills a fascination with history in students, including a requirement to study Latin. One graduate, Jared Diamond, wrote a widely acclaimed book on history, *Guns, Germs, and Steel*, which set out to explain how the evolution of power and technology through history can largely be explained by environmental differences that were reinforced over time by feedback loops. Greg had the opportunity to speak with the distinguished author and gained an even better appreciation for the value of Diamond's approach of taking a broad view of history and using that perspective to gain a deeper understanding of present circumstances and what may unfold in the future. "I think in a way, I was trained to look at things the same way, of taking a big-picture view, of thinking that history matters a lot," said Greg. "And that's the perspective I took going into the solar industry."

Research already existed on solar's historical development, but it tended to be localized to specific countries; nothing existed that attempted to stitch various countries' stories together to gain a more cohesive picture. Greg knew that to do so, he would have to pick the brains of those most intimately involved; that would be the only way to fully explain the technology's evolution. Part of his research—which culminated in a book, *How Solar Energy Became Cheap*—included speaking with seventy-five people in eighteen countries. Each interview began with a standard list of questions, but he also allowed for detours when the interviewee began to get more excited "and started dropping anecdotes about a dinner they had in China with lots of alcohol involved and money moving around and that explains how they started their company." There were many factors he identified through the process that, like the influence of alcohol, are difficult to put into an analytical model: the influence of countries' differences—culture, education, financial markets—or the impact of people moving around the world and dispersing intellectual property and equipment in the process.

The commercialization of solar technology had its beginnings in the U.S. in the 1950s at Bell Labs, which had been founded in 1925 to research technologies that could not only serve the burgeoning telephone market but might also lead to socially beneficial applications that would help Bell Telephone preserve its monopoly. (It was also the same Bell Labs where Peter Eisenberger of Global Thermostat later got his start as an experimental physicist.) Researchers at Bell Labs were given broad freedom to pursue their curiosities to investigate possibilities for new technologies that did not have immediate applications. One researcher studying the ability of silicon to transmit a radio signal discovered in the process that silicon was able to facilitate the flow of electricity when exposed to light. Through subsequent trial and error, the efficiency of converting sunlight to electricity with photovoltaic technology was eventually increased from less than 1 percent to 9 percent—enough that the U.S. military took interest in using the technology to power remote applications. Oil embargoes in the 1970s catalyzed more investment in solar energy, and costs fell by a factor of five as the U.S. led worldwide production until the 1980s. Funding for additional solar research in the U.S. wound down in the early 1990s though as increasingly fierce global competition pushed research and development funding into less risky opportunities with more immediate payoffs.

At the same time, research funding was ramping up in Japan, which leveraged its top-down industrial policy to stimulate innovation and exploit niche markets for solar technology, beginning with ocean navigation lighting. But labor costs were high in Japan, quality was overemphasized, and they overinvested in one specific type of photovoltaic generation. As a result, Germany was able to surprise Japan and surpass it as the leading manufacturer in 2007, instituting huge subsidies for suppliers investing heavily in automated processes that enabled large-scale, low-cost production. China in turn took everything done up until that point, put it together in different ways and then applied far greater ambition than anyone else to scale up and take the lead in 2009. The country is now responsible for nearly two-thirds of worldwide solar panel manufacturing, but it was the distinct characteristics of all countries who took a turn leading the commercialization of solar energy that led to that outcome.

Individuals also played a pivotal role in commercializing solar, including often overlooked people pushing solar energy along. "I think a lot of people dismiss early adopters," said Greg. "Maybe they're dismissed because people think of them as just hippies, or they were dismissed from a systemic perspective because it's too tiny to make any difference. But that's what really kept the industry alive in the U.S. in the 1980s and 1990s, were these small installations."

He also likes to highlight the role of two individuals who played an outsized role. One is Martin Green, a professor at the University of New South Wales in Australia, who is often referred to as the "father of photovoltaics." When the 1970s oil crisis occurred, Green was an Australian university graduate with an interest in researching the potential of solar energy. While completing a PhD in Canada, Green visited the U.S. and spent time with the twelve different companies that were manufacturing solar panels at the time to learn as much as he could about all aspects of their manufacturing processes. When Green returned to Australia and began teaching at the University of New South Wales, he started up a research group that used equipment discarded by the U.S. companies he had visited to find greater efficiencies in converting solar energy to electricity. The group set an efficiency record for solar photovoltaics of 18 percent in 1983, followed by a three-decade string of more records: 20 percent in 1989, 24 percent in 1994, 25 percent in 2008 and 40 percent in 2015.

Shortly after Green started up the photovoltaic research group, China began sending students overseas to receive training in various industries, gaining expertise that could eventually be used in China, it was hoped. An ambitious exchange student named Shi Zhengrong knocked on Green's door one day seeking full-time work. Shi had managed to complete a university degree in China after being given up for adoption at a young age by parents who were poor farmers struggling to recover from China's great famine. Green offered him a scholarship to complete solar-cell research, and the ambitious new student finished a PhD in only two and a half years, finding efficiencies in photovoltaics along the way. Returning to China hadn't appealed to Shi up until that point as he didn't think the infrastructure was there to facilitate starting up a company, so he accepted a position overseeing research for a company Green spun off from the university that aimed to develop next-generation technology. When the Chinese market eventually started growing, Shi's unrelenting drive led him to a local government that was keen on his idea. They pressured some unrelated companies to invest, and a new company led by Shi, Suntech, began producing panels in 2000. When the German market expanded significantly a few years later, Suntech was able to leverage its low-cost production, and by 2009 it had become the largest manufacturer of solar panels in the world.[6]

Shi's company was able to navigate around one potentially significant impediment to quickly scaling manufacturing: the increased risk of supply chain constraints. As manufacturing

---

6 Suntech was a victim of its own success, as numerous local government officials in China tried to replicate its rapid growth—there were 123 manufacturers of solar panels in China by the end of 2010—leading Suntech to declare bankruptcy in 2013.

of solar panels ramped up, its supply chain was tested the most when a surge in German demand led to a significant increase in production. Until that point, manufacturing had benefited from using widely available feedstocks, mainly sand used for silicon and glass, which not only drove costs down but also helped to avoid resource constraints. And given that silicon semiconductors are found in most modern technology, there were enough silicon scraps just from computer processors to satisfy the lower purity needs of the solar industry. But then the production of panels accelerated and supply chain constraints materialized in late 2007, causing the price of silicon to skyrocket from $25 per kilogram to over $400. Solar manufacturers found ways around the constraints, including integrating silicon production into their own manufacturing. Shi insourced supply, finding a local source that was reliable and inexpensive. To train his employees to become more efficient at producing silicon, he sent them to his University of New South Wales alma mater, where a former professor, using a simulated production facility, graded them based on the yield they could achieve.

In the foreword to Greg's book on how solar became cheap, Green echoes Greg's views on how decarbonizing technologies analogous to solar can benefit from the solar experience to achieve a meaningful scale in time to be impactful. "Many good things are happening and we are moving in the right direction—we just need to move more quickly," wrote Green. Despite solar technology continuously surpassing expectations, electricity generation from solar photovoltaics still only accounted for about 3 percent of global electricity generation in 2019. "The first commercial solar cell was produced in 1958 and the first time solar was beating fossil fuels without subsidies was 2018 or so, and now that it's cheap, it's going to take a decade or two to scale up and reach its potential," said Greg. "How do we speed things up? How do we accelerate innovation? We need to go from an eighty-year period from commercialization to large-scale deployment to a twenty-year period, so that in the 2030s and 2040s we've got low-cost technologies that can address the climate problem."

What has the power to accelerate development exponentially? Greg refers to it as policy robustness, which allows investors, whether private or public, to have confidence that demand for a product will continue regardless of any policy specific to one government. That was a revelation that came to Greg when he interviewed Silicon Valley investors early in his career to research how they evaluated the risks involved in making investment decisions, including government policies. "I had a complicated approach that I laid out to a venture capitalist, and he just kind of laughed," said Greg. "It's way simpler than that—we just ignore it. What the government giveth, it can take away," replied the investor.

One important factor in China's willingness to allocate significant resources into boosting solar manufacturing was the robustness provided by the breadth of solar policies in consumer countries. That gave China confidence that if one market disappeared, others would remain. When demand incentives were dramatically curtailed in Germany after a change in government in 2007, there was still Spain and Italy and California. And when all those markets disappeared following the 2008 financial crisis, China had made enough progress that it decided to create more demand within its own borders, subsidizing demand internally and becoming the world's largest solar market in the process. China's aggressiveness was also buoyed by belief that demand for climate-related solutions like solar will continue to grow regardless of specific policies.

In trying to answer the question of how best to accelerate commercialization of technologies needed by the world, Greg points to what may be the most useful analogue: vaccine development, which went from a best-case track record of four years from development to deployment, for a mumps vaccine, to under a year for COVID-19 vaccines. A critical factor in encouraging companies to dramatically accelerate investments in rolling out vaccines were advance purchase agreements: commitments to spend public money before a product has been developed. In the case of COVID-19 vaccines, nearly $100 billion worth of these agreements were signed in 2020, most of them with small and medium enterprises as opposed to big pharma. Scientists had benefited from decades of research that contributed to the development of the vaccines, and thousands of volunteers offered up their deltoids to expedite the clinical trial process, but it was the elimination of commercialization risk by governments that was arguably the most critical factor in shrinking the timeline for such a massive rollout.

Can these lessons be applied to direct air capture? Greg believes so. Following the solar timeline, direct air capture would be low cost by 2077 and widely adopted by 2100; much too late in his opinion. What if instead it was low cost by 2040 and widely adopted by 2060? Or sooner? So far, some of the same accelerators that allowed a company like Suntech to scale up production at 120 percent per year are already working in direct air capture's favor. The U.S. has taken the lead in implementing policy (the 45Q tax credit and California Low Carbon Fuel Standard) that has enabled the first large-scale facility, Carbon Engineering's DAC 1, to begin construction, and other markets might eventually have high enough carbon prices to help build a carbon removal market. The number of companies actively developing the technology remains small, but they are testing new designs, spreading knowledge, training staff, borrowing knowledge from other industries and disrupting manufacturing processes to improve cost and performance.

For most products and services, there is a learning rate effect whereby production costs decrease as volume increases. One key aspect of solar's evolution has been that relatively small unit sizes have amplified the learning rate effect. To date less than a thousand nuclear reactors have been built, compared to about three billion solar panels. "That's a few million times more chances to improve than nuclear, to incorporate some new production technology or design of the device," said Greg. The small unit size allows for the opportunity to take advantage of niche markets at the outset, like powering small electronic devices, but does not preclude larger projects from eventually being built, like China's recent announcement of an enormous three-gigawatt project.

Greg was trying to better understand what the learning rate effect for direct air capture might be when he began looking into the technology in its nascent stage in 2009. The more research he did, the more he kept seeing Klaus Lackner's influence. He invited Klaus to the University of Wisconsin to give a lecture on the subject, and the two were able to spend the remainder of the day talking about the possibilities for scaling up direct air capture, including what the technology's rate of learning might look like. At the time, Greg thought it might be analogous to pollution control equipment, like sulfur dioxide scrubbers, which implied that for every doubling of output, the cost might drop by about 10 percent. Klaus was more optimistic. Direct air capture technology might not be of small enough unit sizes to benefit from the same number of iterations in the design process that came from producing solar panels, but he thinks that for every doubling of output of DAC units, the cost of manufacturing might drop by 20 percent—slightly lower than solar's rate of 24 percent and slightly higher than wind's, 15 percent.

As an example, Climeworks's approach of using more modular units, which are more akin to wind turbines than solar panels in terms of unit size, are small enough to allow for the greater innovation that comes from the ability to constantly tweak the manufacturing process to eke out more efficiencies. And Carbon Engineering's larger-scale approach does not preclude them from reaping some of the same efficiency gains, as many of the components in their process can be modularized. Some quick and dirty math shows the impact of a 20 percent learning rate: if one assumes direct air capture costs $500 per ton, Klaus's estimate of Climeworks's cost for the first commercial plant, the capture rate would only have to grow a thousandfold, from kilotons to megatons, to get to $50 per ton. Or, you could just compare that theoretical tenfold reduction in cost to the 10,000-fold reduction in the cost of solar energy.

Comparing the two technologies is not accepted by everyone. Howard Herzog, for one, is still pessimistic about the prospects for direct air capture. Rather than making a comparison to a technology that has been successfully commercialized, like solar, he suggests looking at one that never got off the ground. The Kemper power project was a coal gasification project located in Mississippi that was touted as paving the way to supplying clean electricity from coal. Ultimately, though, the cost ballooned from $2.4 billion to $7.5 billion, and the project was abandoned after operating for only about a hundred hours. A litany of factors have been blamed by the company developing it—poor market prices, poor weather, generally poor luck—and by journalists who have reviewed internal documents—poor design, poor construction, generally poor management. Inevitably there will also be hiccups as direct air capture plants are rolled out. Regardless, the first commercial direct air capture plant has already beaten Herzog's $1,000-per-ton cost estimate (he has now lowered his estimate to a range of $600 to $1,000 per ton by 2030).

Direct air capture also has one key advantage over solar, which is a lack of competition. Other carbon removal options are competitors, in a sense, but given the scale of the challenge, and the need for as many potential solutions as possible, carbon removal opportunities are all effectively on the same team. That wasn't the case for solar energy, which has faced opposition from fossil fuel industries as a competitor and has been grossly outmatched in lobbying efforts to influence public policy. Similar vested interests who might try to impede the progress of direct air capture are less apparent—it's a large pie to slice up. In 2010,

Greg wrote a paper with a colleague that examined who would have the most to gain from developing direct air capture technology. They concluded that Saudi Arabia had the greatest incentive to fund research and bring the costs down because of the significant impact that would likely have on the value of their oil reserves. "From what I can tell, they haven't read our paper or listened to what we said they should do ten years ago," said Greg.

The more likely sources of resistance to a massive direct air capture buildout would be concern with how public funds are invested and with the environmental footprint of the technology. To address that risk properly will be a major undertaking. To do so, Greg believes the research community, himself included, needs to find innovative approaches to earn public acceptance of the significant infrastructure that will be required to reach large-scale adoption in a timely fashion. "How do you go from a couple of pilot plants today, to say millions of tons by the end of the decade, to billions of tons by 2050?" asks Greg. "There are a lot of ways this could be done badly." To avoid any critical failures, he believes it's crucial to understand both the potential for social impacts and ways to generate support, and so he's part of an international team of academics interviewing people in thirteen countries to help guide research and support policy implementation.

The single most important factor in achieving what Greg envisions, in the absence of multibillion-dollar prepurchase agreements or high enough carbon pricing, is having a market to sell to and not being too picky about what markets look like at the outset. "That's how these technologies get started. You have to find some initial markets to get it going and then you can focus on where it has the most social and environmental benefit," said Greg. "Making sure there are customers is what really leads to the investment and the scale-up over time."

CUSTOMER FEEDBACK

ONE CARBON REMOVAL CUSTOMER is trying to send a message to the entire sector: "We're coming!" To reinforce the point, Lucas Joppa, Microsoft's first chief environmental officer, is working hard to put money on the table. "One of the things I was quite adamant about when we started talking about what Microsoft's carbon efforts need to be was carbon removal," Lucas said. "That's been kind of a taboo subject in the environmental community for a long time because there's a bit of a moral hazard there. But if you read the IPCC reports, it becomes clear: there is no economically feasible scenario where the world stays under a 1.5-degree future and doesn't significantly overshoot its carbon budget and have to remove it."

In 2020, Microsoft announced it would go carbon-negative by 2030 and remove enough carbon dioxide from the atmosphere by 2050 to offset all of the company's historical emissions. "I thought we were going to get a lot of pushback," said Lucas of the announcement. "When you're in the sustainability space, no matter what you do, people complain about it. And we didn't get that many complaints. I was shocked." But the positive feedback was still troubling. "The number one thing that surprised me was everyone talking about how ambitious Microsoft's climate commitments are. And that worries me, because our climate commitments are perfectly in line with what the best available science says everybody has to do."

Microsoft has become a climate leader not by benchmarking their climate goals against what other large companies are doing but because of a commitment to trusting the science. Lucas epitomizes the company's approach: he had a PhD in conservation biology when Microsoft hired him to write scientific papers for the company's blue-sky research arm. One of his main interests was exploring how computing and machine learning could help illuminate the intricacies of the relationship between humans and the ecosystems we rely on. He wrote a memo in 2017 arguing that the company's artificial intelligence research should focus on areas that would support the often-underfunded organizations tasked with solving the world's environmental challenges, providing them with better information about agriculture, biodiversity, water and climate change. The memo led to AI for Earth, a five-year, $50-million investment with 450 projects in over seventy countries. Its circulation also led to Lucas being appointed Microsoft's first chief environmental officer.

His new role coincided with a significant increase in Microsoft's environmental ambitions. The company had announced in 2012 that it would be carbon-neutral by the following year but now recognizes the shortcomings of that hasty proclamation. For one, they had only attempted to offset emissions associated directly with Microsoft operations such as data centers, software development labs and office buildings. In total, those Scope 1 emissions were about 100,000 tons per year, and Scope 2 emissions, such as from the electricity and heat used in Microsoft buildings, were about four million tons. Excluded from the calculation were the much larger Scope 3 emissions—all indirect emissions from employees, suppliers and customers—which totaled twelve million tons annually.

The other shortcoming that Microsoft identified with their initial approach was that they were paying to avoid negative impacts, like cutting down trees for example, instead of paying for a positive impact, like planting more trees. From an offset perspective, protecting a forest may result in carbon removal as compared to a hypothetical situation in which those trees are harvested. But the forest might not actually be at risk of being logged. Or protecting one forest might just lead to the harvest of another forest instead. Or perhaps the forest would have been protected anyway under a conservation program. In any case, there's no net benefit. To truly offset emissions while not being dependent on the actions of others, new trees would need to have been planted to remove carbon dioxide from the atmosphere.

With newfound ambition, in 2020 Microsoft launched its plan to become truly carbon-negative by 2030 and remove enough carbon dioxide by 2050 to offset all historical emissions since the company was founded in 1975, which would amount to twenty-four million tons.[7] Emissions will be curtailed mostly by shifting entirely to renewable energy for data centers and buildings by 2025 and electric vehicles by 2030. To reduce the more unmanageable Scope 3 emissions, the company will expand an internal carbon-pricing scheme to incentivize all suppliers to reduce their own emissions by attaching a weighted carbon

---

7 Microsoft is also putting money on the table as an investor. "If we as a global society are going to achieve a net zero economy by 2050 then some of us who have more—have more resources, have more technology, have more motivation—are going to need to do more," said Lucas. In 2020, the company announced a billion-dollar Climate Innovation Fund that will be deployed over four years to invest in technologies like direct air capture that need to be scaled quickly. From Lucas's perspective, it's an underfunded market along the risk-return spectrum for capital being allocated to climate action—somewhere between less risky markets like renewable energy and moonshot investments.

price to their bids. That would still leave an estimated six million tons per year from difficult-to-mitigate sources like air travel and less controllable Scope 3 sources.

To offset the combination of historical and future hard-to-abate emissions, Microsoft began purchasing carbon removal services in 2021, selecting fifteen different projects out of 189 applications. The vast majority of carbon removal came from projects with relatively short lives, mostly tied to enhancing the ability of forests to be a carbon sink. At the shorter end of the durability spectrum, for example, Microsoft is restoring historically dense forests in India by paying farmers to replant degraded areas (farmers receive 70 percent of the profits).

Forestry projects feature many ancillary benefits, such as improving water resources and protecting endangered species, but they also have shortcomings, including unpredictability. That was a painful lesson for Microsoft when another project, the Klamath forest in Oregon, which represented nearly a fifth of the total carbon removal volume they had purchased, erupted in flames in 2021. The area had been overharvested for many years, with the result that it represented a much smaller carbon sink than it would have otherwise. Through carbon removal payments from Microsoft, the forestry company that owns the lands committed to accelerating management practices to improve the forest's carbon uptake. Their main tool, paradoxically, would be removing more trees to improve forest health and reduce the forest fire hazard. Unfortunately, half a year after Microsoft purchased carbon removal associated with the Klamath forest, it was engulfed in the 400,000-acre Bootleg Fire. "We've bought forest offsets that are now burning," said Microsoft's carbon program manager, Elizabeth Willmott. She remains undeterred though: "We don't want this to force us to pull out of investing in nature-based solutions." Instead, she suggested that buyers must get better at understanding the risks. That's part of the reason Microsoft is ramping up the purchase of carbon removal now, so that they can comfortably hit their 2030 target with the knowledge of what works and what doesn't.

Given that most of the excess carbon dioxide emitted by human activities will remain in the atmosphere for hundreds or thousands of years without human intervention, carbon removal needs to occur on a longer time scale than the bulk of the projects selected by Microsoft, which might remove carbon for several decades before releasing it back into the atmosphere—if they don't burn down first. To that end, Microsoft has also sought out longer-term solutions.

Three of the projects they funded use biochar to absorb carbon dioxide. The use of biochar dates back thousands of years to the Amazon River basin. Unforested areas were coveted for farming but prone to soil deterioration as frequent rains and floods washed away nutrients. That led farmers to clear areas with fire, and in the process, they discovered a way to improve soil fertility. If cleared trees were burned without the presence of oxygen in a rudimentary kiln, it would create charcoal-like biochar, known as *terra preta* ("dark earth") in the region. The *terra preta* was mixed into the soil along with bone fragments, broken pottery, compost and manure, resulting in a highly fertile soil that was better able to retain nutrients and withstand the leaching that deteriorated other soils in the area. Recently discovered patches of *terra preta* have been said to sustain papaya and mango crops that grow three times faster than on surrounding soil. They've also been found to contain nearly three times as much carbon as unimproved soil. The biochar projects Microsoft purchased carbon removal from would use a different kind of kiln, employing a process called pyrolysis that uses high heat without oxygen present to convert agricultural and forestry waste—which would otherwise decompose and release carbon dioxide—into biochar and gases. The gases are then converted into a bio-oil fuel and the carbon-rich biochar is added to soil, where it enhances plant growth by helping to retain water and nutrients for hundreds of years.

Another project with a much longer-term source of carbon removal began with biochar before evolving into a more novel way of sequestering carbon. An aerospace engineer named Kevin Meissner was looking for impactful problems to solve after leaving the aerospace sector and concluded that carbon removal was an underdeveloped technology with significant potential. He began his quest with plants, given their obvious carbon removal abilities. He eventually zeroed in on biochar and began exploring how to make a business out of selling it, possibly to farmers as a soil additive. When that didn't prove as enticing as hoped, he looked at the pyrolysis process itself: by fine-tuning how the biomass was heated, production of any output—biochar, hydrogen, bio-oil—could be optimized. With no attractive market for biochar, and bio-oil not being a practical fuel given its low energy content and relatively short life as a liquid, the most attractive output seemed to be hydrogen. A small market existed for renewable hydrogen to power vehicles, with potential for other applications. A company was formed called Charm Industrial, money was raised, the team was expanded, a pilot plant was built—but the economics still didn't work. The cost of trucking agricultural waste to a plant and then delivering hydrogen to customers was prohibitive.

To reduce the transportation costs, the company's new chief scientist, a friend lured from the aerospace industry named Shaun Meehan, pushed the company to break up the pyrolysis machine into two modular units. One unit could process agricultural waste on-site into bio-oil, which would be cheaper to transport than the waste, while the other unit could reform the bio-oil into hydrogen or other gases directly on-site for customers like vehicle fueling stations. Before they could test whether they had a viable business plan, though, an unexpected business opportunity materialized. A company named Stripe, which provides payment processing software, announced a plan to purchase $1 million of carbon removal per year. It dawned on

Meehan that instead of refining bio-oil into hydrogen to be sold as a fuel, the bio-oil, which is about 43 percent carbon, could just be injected into disposal wells. Its tendency to harden over time, a disadvantage for use as a fuel, gives it a cost advantage over gaseous carbon dioxide that must be compressed before it can be sequestered underground. Stripe purchased an initial 416 tons of carbon sequestration from Charm at $600 per ton after an extensive review process. Other customers, including Microsoft and Shopify, followed.

Sequestration of carbon from otherwise decomposing agricultural waste could be an effective option for avoiding carbon dioxide emissions, but it does not qualify as carbon removal. It's akin to capturing carbon from smokestacks as opposed to the atmosphere. To be a legitimate form of carbon removal, vegetation would have to be grown specifically for sequestering carbon underground as bio-oil. But growing vegetation specifically for carbon removal creates potential land use issues if done on a large enough scale to be impactful. That was always apparent to David Keith, who was one of the first people to recognize the possibility of combining the generation of electricity from biomass with carbon capture and storage as a carbon removal option. In the late 1990s, while a professor at Carnegie Mellon University, he published papers and gave talks on the potential of what later came to be known as BECCS ("bioenergy with carbon capture and storage") to create a longer-term sink than just planting more trees. He even pushed the IPCC—with some success—to include BECCS in a 2005 special report on carbon capture. But he cautioned at the time that the significant amount of land required for large-scale use of BECCS as a climate mitigation tool would create too many other social and environmental issues, a view he still holds.

It's a potential constraint on other options that entail using nature's built-in capacity to draw down carbon dioxide levels. Tim Flannery, the Australian scientist and author, has investigated many of them since his interest in carbon removal was piqued during the twelve years he spent as a judge for the Virgin Earth Challenge. He is now focused on two options, silicate rocks and seaweed, that he believes are most promising for economically delivering carbon removal on a sufficiently large and long-term scale. "The way I look at the issue is that we have an urgent problem and we need to develop scalable solutions—tens of gigatons per year," he said in a 2021 presentation to the AirMiners, a group of entrepreneurs, researchers and funders seeking to collaborate on carbon removal solutions. To put the amount in perspective, he uses the analogy of carpeting the lower forty-eight U.S. states with trees, which might result in a drawdown of about eight billion tons (or eight gigatons) over the course of a century.

The various processes that have formed and unformed the earth's crust over billions of years—weathering, sedimentation, metamorphism, volcanism—have all involved carbon dioxide. Weathering is the process within the geological cycle where carbonic acid falls as rain and breaks down rocks into dissolved ions. The rock could be carbonate rock, like limestone, or silicate rocks, which make up over 90 percent of the earth's crust. It's a very slow-moving process, taking millions of years. But what if weathering could be accelerated similarly to how Klaus Lackner had theorized in his 1995 paper that explored the possibility of carbonating different types of minerals? One option is using a silicate called olivine, which is very abundant below the earth's surface but is not seen much on the surface, where it weathers quickly. If olivine is brought to the surface, crushed and spread around soil or beaches or other areas, a process referred to as enhanced weathering, one kilogram of the ground-up material could capture 1.25 kilograms of carbon dioxide within months, according to Flannery. "The science that's been done around this suggests that the utilization of silicate rocks can offer us a drawdown potential which is the equivalent of between 30 and 300 parts per million of CO_2 out of the atmosphere by 2100," he said.

Seaweed is another option Flannery finds promising. Seaweed grows much faster than land-based counterparts; it just scrubs carbon dioxide from oceans instead, a repository with about fifty times more actively circulated carbon than the atmosphere (removing carbon dioxide from the oceans allows for more carbon dioxide to be absorbed from the atmosphere). Seaweed's greater productivity for capturing and sequestering carbon dioxide is the result of two factors. One is a faster growth rate. By not having to divert any of the glucose produced by photosynthesis towards non-photosynthesizing appendages—trunks, roots, branches—it can be devoted entirely to more productive parts. As a result, one type of seaweed, giant kelp, can grow up to two feet per day. The other advantage seaweed has is that it can naturally sequester carbon, so long as it drifts to a final resting place about 3,000 feet below the ocean surface. Below that level, carbon from decomposed seaweed is effectively taken out of circulation with the atmosphere, remaining sequestered for a lengthy period. "The carbon embedded in that seaweed stays out of the coupled upper atmosphere-oceanic system for centuries, millennia or maybe millions of years," said Flannery. "The science is still fairly open on that, but it's certainly long-term enough storage to give us some real options for utilizing a pathway that gets CO_2 out of the atmosphere and at the very minimum gives us a break and maybe helps permanently solve the problem."

Seaweed cultivation is an established industry, with farming practices dating back centuries, and also has the benefit of providing enhanced habitat for marine life such as shellfish and oysters. Offshore kelp farming would be much more complex, but it has been tested. During the 1970s energy crisis, the U.S. Navy conducted large-scale kelp farming experiments, with the primary intent of finding an economic source of biofuel. It was envisioned that a 100,000-acre farm could be built at a cost of $2-$4 billion that might produce not only forty million cubic feet per day of methane gas but also enough fertilizer and livestock feed to provide for a city of 500,000 to 800,000 people. Over a twelve-year period, about $20 million was spent on the offshore experiments. Floating reefs, moored just below the ocean surface to avoid disrupting marine traffic, were tested off the coasts of California and New York. Nitrate-rich water was pumped from up to 1,000 feet below to nourish the kelp, which had been fastened to a grid of polypropylene lines by navy frogmen. Despite suffering several weather-related setbacks, the experiments proved that open-water kelp farming was feasible: nutrient-rich water was brought to the surface, growth rates were equivalent to naturally growing giant kelp and reproduction was successful. But when concerns that oil and natural gas might run out fell away in the 1980s, the research was shelved.

Flannery wonders if there are better locations for offshore farms, like the doldrums (known formally as the Intertropical Convergence Zone), the typically windless area encircling the earth about thirty degrees north and south of the equator, depending on global weather patterns. Solar-powered, robot-operated mobile farms floating around the deep ocean there would not interfere with coastlines or fragile ecosystems and would allow for easy jettisoning of the kelp's carbon-containing fronds to the deep ocean floor. Of course, there are unknown impacts to bombarding the deep ocean with vast quantities of seaweed, like how it might interact with deep-sea life, but Flannery thinks there's an important mitigating factor. "The oceans are very, very large. The oceans below three kilometers are the largest biome on earth by far; they represent 95 percent of all habitat. They are so large, in fact, that if we were to take half of the atmospheric CO_2, which would be 200-odd parts per million, and put that into the deep ocean, it would increase the CO_2 concentrations of the deep ocean below three kilometers by just 2 percent."

Flannery ranks the potential of silicate rocks and seaweed for carbon removal ahead of direct air capture because of one reason: cost. Until the cost of DAC comes down, he doesn't see much scalability potential. "I hope it may be [affordable] in the future, but at the moment it isn't." But the costs of implementing large-scale carbon removal using silicates and seaweed are not yet well understood either. Until such time, Microsoft, for one, isn't put off by the current costliness of direct air capture. It purchased 1,400 metric tons of carbon removal from Climeworks, equating to a mere 0.1 percent of the total carbon removal volume purchased by the tech giant in 2021.

"The big corporate buyers have realized the different quality of different approaches to carbon removal," said Jan Wurzbacher of Climeworks. Despite the measly amount relative to Microsoft's total contracted volume, Lucas Joppa is biased towards the technology: "The reason I'm excited about that is I'm very simpleminded and there are simple ways to think about direct air capture: price per ton of removal and price per ton of sequestration. You don't have to account as much for competing land use issues as you do with some other techniques like BECCS and enhanced weathering." Microsoft hasn't disclosed how much they paid for the 1,400 metric tons that Climeworks will sequester below the ground on their behalf, just that it is more than fifty times greater than the cost of most of the nature-based approaches (another volume purchaser, Stripe, paid Climeworks $775 per ton). "The innovation is there. The engineering is there. The scale is not there. And the price is way off," said Lucas.

In 2021, Shopify became the largest purchaser of direct air capture carbon removal, prepurchasing 10,000 metric tons' worth from Carbon Engineering in addition to 5,000 metric tons they had already picked up from Climeworks. "It's remarkable that we're the biggest buyer," said the director of Shopify's Sustainability Fund, Stacy Kauk. She would like to see prepurchase and cooperation agreements stacked up in a nice pile for every facility long before they come online, whether it's Carbon Engineering's million-ton-per-year facility being oversubscribed long before it comes online in 2024 or similar stacks piling up for every other project. "It's not a technology barrier; it's not a location issue; it's not a renewable power problem—it's just the appetite and the capital to get it done. In my mind, that's what has to happen."

When Shopify went public in 2015, numerous filings were provided to securities regulators. In the middle of one lengthy registration filing overflowing with information about the company was a letter from CEO Tobi Lütke. He described how Shopify's first company was his own business. He wanted to sell snowboards online but couldn't find software that wasn't expensive, complex, inflexible; software that wasn't tailored to large businesses transitioning online. The company, which now has over a million customers using its platform to build online businesses, has focused on making it easier and cheaper to become an entrepreneur. At the end of the letter, he laid out his hopes for the company. One aspiration was that Shopify is still around a century from now.

That line stuck with him. What's the point in building a company that will be around a century from now if we've failed to look after the planet in the process? As the company has continued to grow since then, it has continued trying to become a sustainable company, switching to renewable power and purchasing cheap offsets. But is that enough? In 2019, Lütke wrote another letter and posted it online: "We've collectively procrastinated so long that the only way to solve this problem is to get carbon *out* of the air, not just prevent more from going in." But like the friction he encountered when he tried to take his snowboard business online in 2004, which led him to create his own software instead, he saw similar sources of friction in the carbon removal market. He decided he wanted to kickstart the market and hired Stacy to tackle the challenge.

When Stacy studied environmental engineering, she decided that her love for math might help her solve some of the world's environmental problems. Her first job found her doing math while wearing a hard hat and boots, helping a variety of industrial companies design pollution control systems. Her calculations were often not well received though: "Oh, that's expensive. Do we have to do that?" It was difficult to stand in the way. "I started to get very good at finding the bare minimum and the regulatory loopholes to scrape by with," she said. Cutting corners was not for her though, so she moved on to work for one of the regulators, Environment and Climate Change Canada. She spent a decade taking on more gratifying challenges, like developing government strategy and negotiating regulatory frameworks. But it similarly left her yearning to make a greater impact. "I was quickly learning a new thing," she said. "Not only is the environment always a cost, [but] perhaps the government isn't going to solve climate change as expediently as I was starting to realize the climate emergency calls for."

It didn't take her long to feel like she was making a difference after joining Shopify at the beginning of 2020: she was given until September to figure out the carbon removal market and spend $5 million on high-quality carbon removal. The company previously considered offsetting all historical emissions, going back to when its CEO built its first website in 2004 and moved in with his in-laws to save money as he grew the company. But it's not a carbon-intensive company, so the amount would quickly be surpassed. Instead, the mission became using the company's purchasing power to be a demand signal for the product they wanted to see become more available and affordable: long-term carbon removal with huge potential to scale. In a stroke of luck, about a week after starting, Stacy found herself at Columbia University for a two-day event exploring what was needed to create a carbon removal market. She found herself sitting beside her recently hired counterpart at Stripe. "What are we doing?" they said to each other at the outset as they tried to keep pace with the information overload. She received valuable assistance and met key individuals, including Klaus Lackner, who offered a warm welcome: "Hey, what do you need to understand this? How can I help you get started?"

Now, Stacy has swiftly become a mentor herself, enthusiastically offering advice to those who solicit it. "It's been really exciting to see that start to happen more," she said. "I would say there's not a week that goes by that we don't talk to another company that's interested in buying carbon removal." She is keen to share what she has learned and make introductions. But whereas she has a boss who's given her money to spend regardless of the cost, she is usually speaking with people who are trying to convince their bosses why it makes sense to spend $1,000 per ton on carbon removal instead of bargain-priced offsets with questionable effect. Oftentimes her usual sales pitch suffices: "We need others to join us with purchase commitments so we can kickstart the market, scale this technology globally and start reversing climate change." If not, she appeals more to shorter-term self-interest: "When the world evolves and you have to buy high-quality carbon removal, you are going to have to go after a scarce resource, potentially."

Shopify is also aware of its own self-interest. One of the company's business philosophies is that by breaking down barriers to building businesses online, more entrepreneurs are created. "One of the reasons Shopify is taking on climate change is because it's a threat to entrepreneurship. Those that really need to build a business, and build value for themselves, can't focus on thriving when they're just trying to survive, and their lives are just getting harder because they're dealing with those adverse effects from climate change," said Stacy. It's a similar perspective to that of Microsoft's Lucas Joppa, who believes it's not enough to move quickly to reverse climate change; it must be done equitably enough so that economic disparities don't widen further. "If Microsoft achieves its goals but the world doesn't, that's a failure for the world and for Microsoft. If Microsoft and the world achieve their goals by leaving many behind, that's a failure. A sustainable transition to a net zero economy is not going to happen unless it's an inclusive one, unless it's a just transition."

Overall, Lucas remains optimistic that progress will indeed be made. "There's no way that human society doesn't address the climate crisis. You can only argue with physics for so long before your forehead really starts to hurt from slamming it into that brick wall." Stacy is also optimistic that new technologies will play a critical role in dealing with 200 years of burning fossil fuels. She views the challenge as one of waste management. "The way I like to think about it, we have all the treatment systems we need to deal with wastewater and purify drinking water. The air is just another fluid that can benefit from all of these technologies. That's where I see direct air capture coming in—it's time to treat the sewage of the sky."

HYPERION II

THE FIRST TIME ALDOUS HUXLEY visited Los Angeles, in 1926, he didn't stay long; the philosophizing author of forty-seven books found the place to be too unsophisticated. "Its light-hearted people are unaware of war and pestilence or famine or revolution, have never in their safe and still half-empty El Dorado known anything but prosperous peace, contentment, universal acceptance," wrote Huxley. A little over a decade later, finding himself disillusioned by the prospect of another world war, he left his native England for India to seek out a change in scenery. He made a stopover in Los Angeles en route and never left. When he later wrote about his new home, he described it as the world's largest oasis.

His newfound reverence for L.A. was tested one day in 1939, when he strolled along the nearby shore with his wife and another couple. Writing about the experience, he described the blissful surprise of finding a beach empty of children and sunbathers. A little further into the walk, they stumbled upon the cause of their solitude: "At our feet, and as far as the eye could reach in all directions, the sand was covered in small whitish objects, like dead caterpillars. Recognition dawned. The dead caterpillars were made of rubber and had once been contraceptives... The scale was American, the figures astronomical. Ten million saw I at a glance. Ten million emblems and mementos of Modern Love." The condoms had been disgorged by a raw sewage outfall just offshore, detected by the beachgoers' noses. "We turned and made all speed toward the parked car," wrote Huxley.

The once prophylactic sewage had found its way to the beach despite a treatment facility, called Hyperion, having been installed upstream from the outfall in 1925. Before then, raw sewage was discharged directly into the ocean. That was consistent with the prevailing view around the world at the time that the cost of treating sewage was unwarranted. One of the most influential experts in the field was an engineer named Allen Hazen, who advised several U.S. cities on the subject and in 1907 wrote a book, *Clean Water and How to Get It.* In it, he argued that while sewage and other effluent could be treated, the cost was prohibitive in most cases. Rather than keeping pollutants out of freshwater resources, it would be much cheaper to just purify water from contaminated sources. "The discharge of crude sewage from the great majority of cities is not locally objectionable in any way to justify the cost of sewage purification," he wrote.

The courts were supportive, typically deciding in favor of polluting businesses and ignoring the social costs of pollution. In one 1905 court decision, as an example, a judge decided that outpourings from an upstream coal mine that had rendered a Pennsylvania family's water undrinkable, and even useless for watering their garden, was just a consequence of a natural product, coal, being discharged in its natural state. Pollution abatement costs, many judges were led to believe, were cost prohibitive and imposing them would force businesses to shut down, which was too great a cost for society in general. Thus, befouled waterways were deemed to be part of the cost of modern life.

Ironically, the onset of traveling by car contributed to the case for treating sewage. As roads began crisscrossing the country-side and cars slowly began to encroach on less-explored areas, an increasing number of urban dwellers descended on swimming and fishing holes. They were often left aghast at the sight of polluted waterways, and some of the more impassioned outdoor enthusiasts were emboldened to demand action. They were hesitant to use the deterioration of recreational opportunities as a rallying call, so instead they focused on health concerns like cholera and typhoid outbreaks that were killing tens of thousands of people every year.

Back in Los Angeles, disgruntled beachgoers had begun lobbying for a treatment facility. They were joined by the local motion picture industry, which produced *The Film with an Odor*, which showed images of sewage spilling onto streets. Their overtures resonated enough that the Hyperion treatment facility was installed in 1925—a simple, primary screening approach that only removed solids, which were then buried in nearby sand dunes. It wasn't an effective solution—one plant engineer estimated that less than 5 percent of solids were removed—and the pipe conveying the screened sewage to the sea was overwhelmed by increasing volumes and began to disintegrate, resulting in the "orgiastic profusion" described by Huxley in 1939. At the time, more funds were becoming available for sewage treatment throughout the U.S. from Depression-era relief programs, but only a little more than half of the systems included some form of secondary treatment such as additional screenings or settlement ponds.

The postwar era brought a yearning for economic renewal that led to more investment. Proponents of using public funding to protect water resources—the Izaak Walton League of America, the Sierra Club, the Audubon Society—highlighted the opportunity that investing billions in sanitation facilities represented, which they argued would pay for themselves by reducing health costs and economic losses. L.A. voters took heed and voted for secondary treatment at the Hyperion plant, which was completed in 1950. It included innovative uses of the waste, such as processing biosolids to produce odorless fertilizer sold to farmers for ten dollars a ton and capturing methane to use as a fuel source.

In 1953, Aldous Huxley returned to the beach he had fled fourteen years prior: "But instead of being, as one might expect, even more thickly constellated with Malthusian flotsam and unspeakable jetsam, the sands are now clean, the quarantine has been lifted. Children dig, well-basted sun-bathers slowly brown, there is splashing and shouting in the surf. A happy consummation—but one has seen this sort of thing before. The novelty lies, not in the pleasantly commonplace end—people enjoying themselves—but in the fantastically ingenious means whereby that end has been brought about."

It was a fleeting triumph. Growing populations in U.S. cities continued to strain sanitation systems through the 1950s and '60s; an increasing number of waterways became choked with various industrial and municipal wastes, and some showed signs of dying. Huxley himself had acknowledged that environmental victories are in no sense definitive or secure: "The art of living together without turning [places] into a dunghill has been repeatedly discovered."

One catalyst for change was a massive oil spill that occurred near L.A. in 1969, the third-largest in U.S. history after Deepwater Horizon and Exxon Valdez. Pictures of oil-soaked beaches sprinkled with dead and dying birds received prominent media coverage. Then, later that year, oily debris floating on the Cuyahoga River in Cleveland erupted in flames. Pictures of the blazing river were displayed in *Time* magazine, along with a description of the river showing no signs of life, "not even low forms such as leeches and sludge worms that usually thrive on wastes." The Cuyahoga had erupted in flames before—the pictures in *Time* were actually from a 1952 fire—but the story sparked nationwide concern and catalyzed calls for action. The same environmental groups that had previously pushed for increased investment in cleaning up waterways campaigned heavily for federal legislation that would bring greater stringency to regulatory oversight. President Richard Nixon addressed calls for action in his 1970 State of the Union speech: "The great question of the '70s is: Shall we surrender to our surroundings, or shall we make our peace with nature and begin to make reparations for the damage we have done to our air, to our land, and to our water?" The federal Environmental Protection Agency (EPA) was created that year and the Clean Water Act was passed the following year.

Meanwhile, in L.A., the burgeoning population's untreated effluent was being piped farther out to sea to circumvent capacity limits at the Hyperion treatment plant. While pushing raw sewage farther out to sea eradicated outbreaks of dysentery among beachgoers, marine life was disappearing. Its absence was discovered by a high school teacher named Howard Bennett, who swam daily in the bay. He investigated further and found that excessive amounts of ammonia, oil, grease and various heavy metals were showing up in Santa Monica Bay. One morning, when Bennett was entering the water for his daily dip, he was stopped by a man waving his arms and shouting that the water was poisoned. Bennett later tracked the man down and discovered he was a marine scientist named Dr. Rimmon Fay who had spent thousands of hours diving along the sea floor to conduct research. Bennett told him he wanted to take action but was told by Fay that nothing could be done. Undaunted, Bennett rallied enough support to get a public hearing in 1985 and convinced Fay to present his findings: Hyperion's discharges were destroying undersea habitat and voiding areas of marine life.

In arguing their position, city officials largely relied on testimony from a former mining engineer named William Bascom, the director of the Southern California Coastal Water Research Project, which had been established in 1973 to assist with obtaining waivers from the new clean water rules. Bascom, who had previously formed several companies to explore for diamonds and sunken treasure, had spent over a decade ostensibly conducting research and convincing authorities that sewage treatment facilities did not need to be upgraded. At the hearing, he countered Fay's findings by arguing that rather than polluting the water, untreated sewage was instead an excellent food source for fish, and their population had increased substantially. His arguments were thrown out, however, when one of his employees testified that data was doctored to support the bogus fish food argument and that staff were threatened with being fired if they released a study showing high toxicity levels in marine animals. The employee also testified that staff were regularly reminded by Bascom that they were making money from the cities and counties of southern California by producing studies allowing those clients to avoid making large investments in sewage treatment.

The city's mayor, who until that point had chafed at the significant investment required to bring treatment in line with the new standards, reversed course and came out in favor of the necessary upgrades. Shortly thereafter, the City of Los Angeles was fined $625,000 for violating the Clean Water Act standards, the largest-ever fine at the time, and a deadline was set for full secondary wastewater treatment by 1998. By the time voters were asked to approve the necessary bonds to begin financing the upgrades, over 250 newspaper articles had appeared throughout the country bringing attention to ongoing sewage spills and the deteriorating condition of Santa Monica Bay. The Hyperion upgrade received overwhelming support, and it was eventually completed in 1998 at a cost of over $2 billion.

IT'S ESTIMATED THAT ROUGHLY $2 TRILLION has been spent on wastewater treatment in the U.S. since the passage of the 1972 Clean Water Act. Putting a dollar figure on the benefits of treating wastewater is not as straightforward. Joseph Shapiro was a graduate student at MIT when he decided to put together a cost-benefit analysis of the Clean Water Act. "Many influential economics papers study air pollution policy, but far fewer study water pollution policy, and it seemed like an asymmetry that could be fixed," said Joseph. The explanation for the dearth of research became obvious in time: there were millions of water pollution readings that were relatively untouched. "I had encountered a lot of researchers who said, 'Oh, I tried to use them but gave up because it was too hard,' but because of stubbornness or ignorance I kept working on it. It took about ten years to produce the first paper."

To estimate the value people put on the cleanliness of neighboring streams and rivers, Joseph and his co-author, David Keiser, also collected data on housing values along the waterways. After navigating through all the data—and hiring a software programmer to help warehouse it—the researchers had created the most extensive single dataset ever built on water pollution in the U.S. The conclusion produced by the model was that while water pollution had decreased over time, there wasn't a high value put on it: house prices increased by about one-quarter of the value of federal grants directed towards the water treatment plants.

Along with publishing their findings, they reviewed high-level research that had been done on the benefits of clean water and found that studies consistently valued the benefits of wastewater treatment at about half the cost. "Water pollution has declined dramatically, and the Clean Water Act contributed substantially to these declines," said Joseph. "So we were shocked to find that the measured benefit numbers were so low compared to the costs." One of the challenges of putting a value on the benefits of reducing water pollution is that it's difficult to quantify how it has directly improved human well-being. "Many of these studies count little or no benefit of cleaning up rivers, lakes and streams for human health because they assume that if we drink the water, it goes through a separate purification process, and no matter how dirty the water in the river is, it's not going to affect people's health," said Joseph. There are, of course, other benefits from clean water besides the impact on human health, but improvements in human health are usually the most readily available barometer of the effectiveness of pollution control regulations. In looking at the 1990 Clean Air Act amendments, for example, the EPA estimated the benefits of the pollution control—which cost roughly $65 billion per year—at a whopping $1.8 trillion per year in 2020. Effectively the entire benefit comes from avoiding premature deaths, with each extended life valued at $8 million, based on studies of how much individuals would be willing to pay to avoid such an outcome.[8]

If the same willingness-to-pay approach for human health outcomes is extrapolated to other benefits of pollution control, the analysis changes significantly. To illustrate the point, Joseph points to the Exxon Valdez oil spill and discrepancies in estimates placed on the cost of it. One study, which focused on lost economic value, estimated the cost of the spill at $3.8 million, based on the impact of fewer recreational visits to the area of the spill. Another study, which applied a willingness-to-pay approach to simply having clean water, estimated the cost at $4.9 billion, based on surveys of what households would be willing to pay to avoid a similar spill in the future—a thousandfold increase over the assumed economic cost. As a result, Joseph wonders if estimates of the benefits of surface water quality are lower than they should be, as they haven't been able to rely on more easily estimable health costs for justification.

8   The willingness-to-pay analysis used by the EPA reflects an aggregate of what individuals would pay for a small reduction in the risk of premature death. For example, if a reduction in a specific pollutant is estimated to result in a 1/100,000 reduction in the risk of premature death and the average amount an individual is willing to pay to avoid that risk is $80, then the implied value of one avoided premature death is $8 million ($80 x 100,000). The study assumed 230,000 premature deaths were avoided in 2020 and valuing each avoidance at $8 million resulted in $1.8 trillion in benefits.

Despite this uncertainty, Joseph views legislation like the Clean Water Act and Clean Air Act as hugely beneficial for society. "The policies were overwhelmingly popular," said Joseph. "The challenge now is to address the environmental problems not addressed or skipped by environmental policies in the last half century." He is confident climate regulations will eventually prevail, but the how and when are more uncertain. "When the U.S. creates air and water quality problems, the U.S. bears most of the costs. If the U.S. creates greenhouse gas emissions, it bears part of the costs, but the rest of the world bears part of the costs and future generations bear part of the costs," said Joseph. "While the Clean Water Act and Clean Air Act were complex, they didn't require the U.S., the E.U., China, India, Brazil and other countries to negotiate what the law would look like."

THE ICONIC BEACHES OF LOS ANGELES, whose condom-laced unsightliness previously exemplified the ramifications of inadequate sewage handling, are also threatened by climate change. How much the beaches will be impacted depends on how much warming occurs, with estimates of sea level rise ranging between one and ten feet by 2100 (the higher end of the range reflects a scenario in which the Greenland and West Antarctic ice sheets destabilize). Adaptation efforts have been hampered by a decline in the amount of sand transported to the ocean from the naturally eroding banks of rivers. Human intervention of various forms—inland reservoirs and dams, channelized rivers, sand mining—has reduced the supply of sand delivered annually by fourteen rivers to southern California beaches by half. As a result, without further human intervention, half of southern California's beaches could disappear by 2100, assuming a midrange sea level rise.

There are numerous options to mitigate the impacts of rising sea levels: engineered solutions like sea walls; adaptation measures like elevating buildings; and renourishment programs that feed beaches with more sand, delivered from offshore dredging programs or from inland quarries via truckload. Sea walls are a less attractive option as they're a barrier against natural replenishment of beaches, causing them to shrink and likely disappear eventually. But the feasibility of constantly renourishing beaches via barge or truckload is not well understood. Overall, the cost of adapting to rising sea levels has been estimated at $4-$6 billion to respond to increases of three to seven feet. If the temperature increase is limited to 1.5 degrees, however, sea level rise by 2300 could be held below two feet.

Trying to estimate the benefit of limiting climate change for the sake of L.A.'s beaches is somewhat straightforward, as are some other benefits of limiting climate change. "Economists and researchers have been developing methods to value environmental goods since the mid-20th century, and air and water quality have been important settings for developing those methods," says Joseph. "Many people are applying them to climate change and improving those methods. There are excellent papers looking at how climate change affects health, firm productivity and home values, just as there are papers looking at those outcomes for air and water quality." Yet, the greatest benefit of limiting climate change is arguably an inestimable one: avoiding tipping points, which might occur when gradual changes become significant enough to cause larger, abrupt, irreversible changes that lead to further warming and in turn increase the risk of triggering more tipping points. Potential tipping points have been increasingly referenced by the IPCC; the panel's sixth report on the science of climate change identified twelve of them, like the melting of the Greenland and Antarctica ice sheets and thawing permafrost. The panel has said that the increased risk of tipping points occurring "between 1.6°C and 2.6°C above pre-industrial levels was assessed to be disproportionately large."

In 2019, L.A. announced a plan to increase its efforts to be consistent with limiting warming to 1.5 degrees Celsius. The bolder plan (a previous sustainability plan had been announced in 2015) called for a 50 percent reduction in greenhouse gas emissions by 2025 compared with 1990 levels and becoming carbon-neutral by 2050. The largest contribution would come from hastening a transition to clean energy and fully electrifying everything—vehicles, public transit, heating, cooking.

To test the feasibility of transitioning to 100 percent clean energy in a relatively short time span, city officials teamed up with the National Renewable Energy Laboratory, one of the Department of Energy research institutes created by the U.S. government in 1977, with the mission of accelerating the commercialization of renewable energy and energy efficiency technologies. To model L.A.'s transition, an unprecedented effort was undertaken using the institute's "Eagle" supercomputer; over one hundred million simulations were run over the course of several years. Factored into the simulations were a multitude of assumptions, from electricity usage to weather conditions to rooftop capacity for solar panels. The study team, which included nearly a hundred people, concluded that the city could get to 98 percent clean energy by 2030 and 100 percent by 2035, all without causing blackouts or disrupting the economy. "The top scientists in the world have taken this from dreamland to reality," said L.A.'s mayor, Eric Garcetti. "To me, accelerating the target date to 2035 increases the benefits. Even if it costs more upfront, we know the long-term costs of living cheaply in the short term are too much to bear."

One obstacle to electrifying pretty much everything in L.A. with clean energy is overcoming resistance. When officials with the Port of Los Angeles reached out to community residents to ask for feedback on whether they should require trucks to be powered by electricity or natural gas, the fossil fuel industry intervened. An investigation by the *Los Angeles Times*, *Floodlight* and the *Guardian* found that several people were hired by a company to appear at the public meetings under the veil of concerned citizens to lobby for natural gas-powered vehicles (participants at the meeting were left to voluntarily disclose whether they were being paid to be there). The company, it was later discovered, had been paid by another company that owned natural gas fueling stations. The logo of L.A.'s largest gas distribution utility, Southern California Gas Company (typically referred to as SoCalGas), also the largest in the country, had appeared on some of the marketing materials. (SoCalGas did not respond to questions from journalists about whether it was involved in the campaign.) Ultimately, Port of Los Angeles officials decided in favor of natural gas trucks, citing community support as a factor.

A couple of years later it emerged that SoCalGas had paid consultants—with funds from regulated ratepayers—to create a nonprofit called Californians for Balanced Energy Solutions. In a practice sometimes called astroturfing, the group was ostensibly advocating for concerned citizens but was really speaking on behalf of the investor-owned utility. In this case, the objective was to sway state regulators against weaning California off natural gas. While it is common practice for utilities to try to exert influence on regulatory outcomes, utilities cannot use ratepayer funds to lobby on behalf of investors. SoCalGas has since agreed to charge its shareholders for any funding of Californians for Balanced Energy Solutions, and the utility also dropped a lawsuit brought against the California Energy Commission for not sufficiently considering the benefits of natural gas.

SoCalGas now appears to be contemplating the transition to clean fuels more fully. A 2021 report released by the company, entitled "The Role of Clean Fuel," states how clean fuels could support electrification paths similar to those modeled by the National Renewable Energy Laboratory. None of those fuels—biogas, hydrogen, synthetic natural gas—is a silver bullet. Biogas, typically a captured waste gas, is hampered by its limited supply: the entire amount available in the U.S. is equivalent to about a quarter of the amount of gas used by California's buildings alone. In contrast, the potential supply of hydrogen is vastly larger. There are different ways of producing hydrogen and the report highlighted two: blue hydrogen, which is produced from fossil fuels like natural gas using carbon capture and storage; and green hydrogen, which is produced using renewable energy. One potential problem with using natural gas to produce hydrogen is fugitive methane emissions during the extraction and processing of natural gas, which could increase with the use of additional natural gas as an energy source for carbon capture and storage. A 2021 study by researchers at Cornell and Stanford universities found that blue hydrogen results in more greenhouse gas emissions than from simply burning natural gas. Green hydrogen, while not yet a mature technology, uses renewable energy to produce hydrogen by splitting water into hydrogen and oxygen and could fuel power generation that would serve as a backup to renewable energy fed directly into the grid, among other possibilities.

Another clean fuel option raised by SoCalGas was traditional natural gas combustion offset by direct air capture. SoCalGas has invested in a DAC company that uses a technology conceived by a U.S. national laboratory, Pacific Northwest National Laboratory, and is being developed by a company called Avnos Inc. It not only removes carbon dioxide from the air but also pulls out water that can be used for irrigation. "We're confident that the demonstration testing of this breakthrough technology will prove what our modeling indicates—collecting significant amounts of water while pulling carbon dioxide from the air results in the most flexible and cost-effective solution in the DAC market," said the CEO of Avnos, Will Kain.

Regardless of what decarbonization path L.A. takes, carbon removal of some form is virtually guaranteed to play a role in achieving carbon neutrality by 2050. "Even with today's best strategies and technologies, there are likely to be residual emissions in 2050, approximately 8.5 percent of our emissions today from sources such as air and sea travel and industry energy use," states the city's decarbonization blueprint. "New technologies will be needed, as well as carbon-negative projects, such as urban forests, to potentially offset carbon emissions."

BEFORE JIM MCDERMOTT MOVED from New York to Los Angeles and started up Stamps.com, one of the financings he worked on was a sewage bond to fund the multibillion-dollar Hyperion modernization. L.A. residents had been asked to vote for the bonds, which would be paid back over thirty years and increase monthly sewage bills from an average of $8 per month to an estimated $18 when payments to bondholders began. Without the bonds, the upgrades would have to be paid directly from revenue and monthly bills would jump to $27. From the city's perspective, it was an easy decision, and they pleaded with Angelenos in a *Los Angeles Times* editorial to vote in favor of the bonds: "Losing this bond issue would only mean more delay, rapidly escalating sewer fees and much higher total construction cost, for the city has no choice but to proceed with the work under court edict."

Municipal bonds provide cities with a cost-effective way of financing capital projects with a social benefit—schools, highways, sewage treatment—by offering a lower interest payment that isn't subject to income taxes. Jim believes carbon removal should also be eligible for funding through municipal bonds. "Dealing with carbon is a fundamental matter. It's a natural progression that regulations are changed to include carbon and the cost is subsidized and spread across society. We have to." In L.A.'s case, the outcome of municipal bond funding could be spreading out the cost of a hypothetical air treatment plant, a Hyperion II for the sky likely located somewhere outside of the city. It could also be gas utilities building direct air capture plants and putting them in their rate base so that ratepayers pay an extra few cents per unit of natural gas.

Regulatory policies that regulate greenhouse gases similarly to the Clean Water Act and Clean Air Act would help make Hyperion II a reality. Whether it's putting a price on carbon or mandating zero emissions, other states might follow California's lead—each with their own version, more cleverly designed than any other state's—until there is sufficient critical mass for the various regulatory approaches to be homogenized at the federal level. "What happens in California is often followed by the rest of the country," said Jim. The leader of one of the organizations that pushed for L.A.'s clean energy plan goes a step further. "Where L.A. goes, so goes the state, so goes the country," Jasmin Vargas of Food & Water Watch told the *Washington Post*. "L.A. and what we do here is being seen as a model. Considering the climate crisis, we have to put extra pressure and focus on doing it right in L.A. or else they'll do it wrong everywhere else."

THE PRIZE

WHEN SPACEX'S FIRST STARSHIP launches into orbit, the reusable, cargo-carrying spacecraft, with a payload of 100–200 tons, will be the largest flying object to ever leave Earth. The ship will be capable of both intercontinental missions—flying to the other side of Earth in forty-five minutes—and interplanetary missions—flying to Mars in four months. The latter will allow humans to begin colonizing Mars, with the ultimate goal of building a self-sustaining city there by 2050. Elon Musk, who founded SpaceX and continues to lead it, believes that humans must become a multiplanetary species because there are too many risks to rely on only one planet—his greatest existential fear is artificial intelligence, which he likens to summoning the demon. One thing the endeavor is not, according to Musk, is an escape hatch for rich people. "Going to Mars reads like that ad for Shackleton going to the Antarctic: it's dangerous, it's uncomfortable, it's a long journey, you might not come back alive—but it's a glorious adventure and it will be an amazing experience," said Musk. "Honestly, a bunch of people probably will die in the beginning. It's tough sledding over there. We're not going to make anyone go. It's volunteers only."

He would also like to reassure observers that space exploration does not make him a hypocrite with respect to his other endeavors, like Tesla's mission to accelerate the transition to sustainable energy. "We have a long-term plan for sustainability of even rocket flights by generating propellant using sustainable energy," said Musk. The starship's propellant will be made up mostly of liquid oxygen derived from separating oxygen from the other gases found in the atmosphere. The other ingredient will be liquid methane created by combining the carbon dioxide extracted in the air separation unit with hydrogen from water. The whole operation will eventually be powered by renewable energy sources to make the propellant sustainable.[9] (On Mars, where carbon dioxide makes up nearly all of the atmosphere and much of the surface is covered by ice underlying a layer of dust, one of the first orders of business will be collecting carbon dioxide and water to create the required methane and oxygen.)

To find more practical carbon removal solutions for the larger task of drawing down atmospheric levels for the benefit of those wishing to remain on Earth, the Musk Foundation is funding a $100-million XPRIZE, the largest incentive prize in history. Announced in February 2021, it will be awarded to whoever can come up with best-in-class carbon removal solutions. More specifically, a $50-million grand prize will be awarded in 2025 to the lowest-cost solution, so long as it can demonstrate the capacity to remove at least 1,000 metric tons per year and

---

9 A greater sustainability concern might be black carbon (or soot). A 2022 study found that the climate impact of black carbon emissions from an increasing number of space flights could be substantial owing to the magnified effect of depositing black carbon in the stratosphere, where it remains for up to four years. The associated warming effect was found to be nearly 500 times greater than from other sources of black carbon.

provide a credible plan for eventually scaling up to billions of tons per year. And, in an evolution from the Virgin Earth Challenge, milestone payments of $1 million were awarded in 2022 to fifteen teams from nine countries pursuing various types of carbon removal.[10] Entrants looking for further assurance that prize money will materialize might also be comforted by Musk's net worth, which is about fifty times greater than Richard Branson's. "I'm open to increasing the prize size over time too. If it turns out somebody really kicked ass and somehow there's not a prize for them, I'll add some more to the prize," said Musk at the launch of the competition. "I don't want somebody to have spent massive blood, sweat and tears; have done something useful; and then get nothing for it. That would be pretty bad."

Overseeing the competition is a physicist named Marcius Extavour. When Marcius (pronounced Marcus) was a kid, his parents would send him off to day camps where he was given materials and a timeline to design something that would then be tested. *This is the greatest thing ever conceived*, Marcius recalls thinking at the time. He enjoyed it so much, and saw so much potential for building problem-solving skills, that he later taught similar workshops for young kids: "I saw how competitions could really drive a lot of collaboration and fun and engagement in a topic; way more than other styles can." Marcius studied engineering as an undergraduate student and got a job optimizing solar panel components in a lab for Nortel Networks before the company imploded. That experience opened his eyes to energy and climate matters, and when he returned to university to complete a PhD specializing in quantum physics,

---

10 The fifteen milestone winners, who were awarded $15 million in total, were selected by a panel of seventy expert judges from 1,133 submissions that were narrowed to 287 teams meeting the eligibility criteria (all 1,133 teams who entered are still eligible for the $50 million grand prize in 2025). In addition, $5 million was awarded in 2021 to twenty-three student-led teams and $30 million will be awarded to three or more grand prize runners-up.

he attended climate seminars in the evenings to better understand the science of both the problem and potential solutions. After graduate school he looked for ways to combine his interest in energy and science with opportunities that could influence government policy.

When the chance arose to join XPRIZE in 2015, Marcius saw it as the perfect opportunity to combine his passions. The XPRIZE Foundation began with the Ansari XPrize for suborbital spaceflight competition in 1994, initiated by Peter Diamandis, an enterprising engineer, physician and entrepreneur. Since that prize was awarded in 2004, more than twenty XPRIZE competitions have been held; the prize money has been awarded in most cases, unless the competition is still active. The first prize Marcius oversaw was a competition seeking to incentivize the conversion of carbon dioxide emissions into useful products. He was skeptical much would come of the prize, because the energy sector's innovation impediments—sunk costs, capital assets, regulation, public safety—might get in the way. "It's hard to make a change and there are arguably good reasons not to make changes... that inertia is a block to change and it's a problem. Could a prize break that up?"

That XPRIZE competition was funded by two energy companies: NRG, a large U.S.-based electric utility, and COSIA, an alliance of companies developing Canada's oil sands. Finalists were given the opportunity to demonstrate that their concepts worked by converting carbon dioxide emissions from a coal-fired plant in Wyoming or a natural gas–fired plant in Alberta. The winner was selected based on the percentage of carbon dioxide in the flue gas converted to a useful product and the value of that product. Both winners, CarbonBuilt and CarbonCure, reduced the amount of cement, an emissions-intensive product, used in making concrete (carbon dioxide is injected during concrete mixing, where it mineralizes into calcium carbonate, a substitute for cement).

As the competition entered its final phase in 2020, the XPRIZE team gathered for what they refer to as an annual ideation festival, an opportunity to discuss what was working and what wasn't. Exponential cost reductions in generating energy with solar and wind since the competition was initiated in 2015 threatened the long-term viability of coal- and natural gas-powered power plants and reduced the potential impact of technologies converting carbon from point-source emissions. There would still be plenty of carbon dioxide available from other industrial plants—cement, steel, glass—with even higher carbon dioxide concentrations, but they wondered if there was an opportunity for an even more impactful climate intervention. What if the Virgin Earth Challenge was too early, and an opportunity remained to hasten the development of carbon removal options?

The team spent a year researching the potential costs and benefits of carbon removal and how best to design a competition, which included seeking feedback from those who oversaw the Virgin Earth Challenge. To fund the prize, a goal was set of raising $25 million from six corporations representing different industries to bring different perspectives together and deliver a message that carbon removal is an industry-agnostic solution. Normally it would take a couple of years to raise that amount of money, but they were encouraged by the reception. When they fundraised for the 2015 carbon conversion competition, a typical response might have been: "Carbon what? Utilization what?" Whereas in 2020 it was more like: "Oh yes, I've heard of that. Let's talk about how to do it and how to do it right." But the COVID-19 pandemic slowed progress, so it was a welcomed boost when Elon Musk agreed to step up and fund the entire competition through his Musk Foundation.

When XPRIZE Carbon Removal was announced in early 2021, it wasn't universally acclaimed. "I think I understand the skepticism," said Marcius. "It's quite difficult to design a good prize.

Or the reverse: it's easy to design a shit prize." The process goes beyond writing down all the ideal characteristics of a solution to a problem and announcing a prize will be awarded to whoever meets them. The key to designing a good prize, according to Marcius, is finding a sector where there are identifiable blockages to innovation. "Of those blockages, are there one or two or three key things that, if addressed, could open up that blockage? If so, then you can probably design a good prize around it." In the case of carbon removal, the blockages were very apparent: a massive scale required (i.e., billions of tons per year by 2050), a lack of currently cost-effective options to get there and a lack of structural support from government and markets.

Just as design competitions have incentivized technological innovation over time, they themselves have also evolved. No longer is it a case of awarding the production of a specific device or functionality; modern prizes attempt to inspire something broader. "Not a black box that's just meant to generate a technology innovation but a prize as something that generates technological innovation and also uses the competition model as a leverage platform to try to drive more attention, more partnership, more collaboration, more scrutiny and more public discussion of the topic," said Marcius.

Of those side effects, an easy one to measure might be the attention generated. The video showing Elon Musk introducing the prize, for example, was watched by over 750,000 viewers within six months of being released. With respect to collaboration, XPRIZE has provided several venues to bring together participants looking for more puzzle pieces with those who have pieces to offer. A secondary effect more difficult to measure is the leveraged investment, but the XPRIZE team has nonetheless tried. They estimated that the $20 million NRG COSIA Carbon XPRIZE generated investments in participating companies of over $200 million.

When Marcius was working on designing the carbon removal competition, he reached out to David Keith for advice. The Virgin

Earth Challenge helped David attract attention and raise funds when he was starting up Carbon Engineering in 2009, but it also diverted resources and focus within the company and competitors. By the time Virgin punted, as David puts it, total funding for direct air capture was much greater than $25 million, so the ultimate effect of terminating the prize didn't amount to much. From his perspective, the original incentive prizes had suitable goals: the Longitude Prize was easy to measure objectively, as was the Ansari XPrize, which required two spaceflights within two weeks using the same spacecraft. In comparison, David struggled with a broad carbon removal prize that, if attempting to find the most cost-effective, environmentally sustainable solutions once they're scaled up, would necessarily be subjective in nature. No truly objective measurement of the long-term cost and sustainability of proposed solutions would be available until the sector had already scaled up, at which point a prize would no longer be useful. As a result, David argued the result would be a political beauty contest.

Marcius acknowledges that entrants will have to convince judges that their cost and scalability assumptions are reasonable, but someone with money is going to have to make a bet on those assumptions at some point. And if XPRIZE does, then governments, investors and customers might be more likely to follow suit. To Marcius, the hardest bar to clear, and greatest counterbalance to arbitrariness, will be the requirement that entrants remove at least 1,000 tons of carbon dioxide per year, which has only been achieved by the Climeworks Orca project so far. "David's right to say it's hard to predict the future, and that's why the demonstration element is so important," said Marcius. "That is the thing that I think protects us from a perverse outcome."

Another concern of David's is the risk that a competition interferes with establishing more practical solutions. As an example, he points to the two companies awarded the previous XPRIZE, which developed technologies that use a fraction

of carbon dioxide emitted from smokestacks as a cement substitute. "I sat around a top government office where people were
very excited about those little companies, and I think it was very
clear that it was standing in the way of the practical ways you
would actually cut CO_2 from cement," said David. "It's allowing government officials and the public to get distracted by little
shiny startups that look like they're going to do something." He
cites a large cement plant near his hometown that emits over a
million tons per year: "We know how to capture from that, we
know where to put it, there's no new technology needed. What's
missing is government will." Which is something David and Marcius agree on: prizes should not interfere with making the most
practical, cost-effective emissions cuts. Otherwise, Marcius has
a message: "Shame on you for being distracted if that's the case.
This is an alternative, but it's clearly not the first option. The first
option is immediate emissions cuts."

AROUND THE TIME THE Virgin Earth Challenge was formally terminated, another carbon removal competition was being proposed
by a U.S. Republican senator from Wyoming named John Bar-
rasso. The bill containing the $35-million competition supported
various climate actions that don't constrain fossil fuel usage, with
awards to be given out to direct air capture projects capturing
over 10,000 tons per year of carbon dioxide. The bill also autho-
rized the Department of Energy to put more money into DAC for
research, development, demonstration and deployment. It was
eventually passed and made into law in late 2020 and has helped
to hasten a shift at the Department of Energy, where funding for
DAC, which had totaled only $11 million to that point, increased
to $68 million in 2020 and 2021 combined.

Noah Deich has played an important role in building bipartisan support for carbon removal. When Noah began his career
working for consulting firms advising large utilities and
energy companies on climate-related matters, the focus was

on reducing emissions. Later, as a graduate student in business, he studied the options—energy efficiency, demand response, renewable energy, battery storage, electric vehicles—that were all needed developments; but it seemed to him they were also moving way too slowly. He was studying at University of California, Berkeley, alongside world-class climate scientists and found himself asking a question he felt they might see as naïve: Do we need to do anything with the carbon that's already in the atmosphere, which will continue to build until emissions stop? The scientists were quite certain that massive carbon removal was indeed required, alongside rapid decarbonization. Yet even though he was interacting regularly with business and policy leaders, this was totally new information. "There was a huge disconnect," said Noah. Eager to learn more about the nascent field, he googled "carbon removal," and the first thing that popped up was the Virgin Earth Challenge. He emailed them to express his interest in learning more about their work, and, to his delight, they responded. He spent a summer working on a consulting project for the competition, where he interacted with Klaus Lackner, David Keith and others. "The Virgin Earth Challenge did an amazing job of bringing together early carbon removal pioneers long before carbon removal had become a core part of the climate solutions toolkit."

Upon graduating, Noah decided there needed to be a home in civil society for figuring out what role carbon removal needs to play separate from any profit motive. He and a fellow student named Giana Amador started up an advocacy group in 2015 that eventually became Carbon180. At the time, he found that most people were not interested in the topic; it was viewed as something for mopping up excess emissions down the road and a distraction from the more serious business of reducing emissions in the meantime. Carbon180 is now located in Washington, D.C., and has seen the recognition of the need for carbon removal grow exponentially, and particularly the need to address legacy emissions as the planet breaches dangerous atmospheric concentration levels. "I think once people start to look at the portfolio of climate removal solutions as an essential piece to solving our climate puzzle, all of a sudden this becomes not an academic exercise but one of possibility and a widely expanded solution set," said Noah.

Carbon180's principal strategy for hastening the commercialization of carbon removal has been to push on what they view as the biggest lever: the U.S. federal government. That has included convincing the government to funnel some of its multibillion-dollar research and development budget towards carbon removal. Carbon180 also played a pivotal role in getting the $50-per-ton 45Q tax credit, which is partly funding Carbon Engineering's first project in Texas, to include various forms of carbon removal, not just projects capturing carbon from smokestacks. The challenge continues to be more about convincing people that carbon removal is a real option than overcoming resistance. "People make handsome profits pulling carbon out of the ground and putting it into the air. So that's naturally a huge political struggle to stop that," said Noah. But so long as carbon removal is done in a way that protects the environment, benefits communities and creates good jobs, there are no interest groups standing in the way. "There's not anyone pushing back, saying, 'This is going to fundamentally ruin our interests as an industry, a community, etc.'" Of course, that will be tested more fully once large-scale projects are actually under construction. As the industry quickly ramps up to the levels Noah would like to see, removing over a billion tons per year by the early 2030s, he expects there will inevitably be political challenges. But the feasibility of scaling up further will be better understood and the process will be fine-tuned. "This is not some exogenous thing that's just going to happen—it can be shaped."

He hopes that carbon removal efforts can create even more political will for climate action, bolstering support for broader decarbonization efforts. "The real question is, how do you do this in a way that doesn't undermine progress towards rapid decarbonization, and in fact, complement that by helping it go faster and creating broader political will for climate action overall?" Part of that will be avoiding the political setbacks that have occurred with solar, as much of the manufacturing has ended up in China. "There's a key role for making sure that early projects are good projects, that they explicitly work to build political will for carbon removal scaling in the future, even if that means not making the cheapest project happen."

KLAUS LACKNER NOW RUNS the Center for Negative Carbon Emissions at Arizona State University. He and Allen Wright (who created Global Research Technologies after leaving Biosphere 2) started up the learning institution and technology accelerator in 2014. Klaus continues teaching classes, tinkering in the lab—or rather his team does most of the tinkering, trying to optimize the various sorbents and other components they're working on— and advocating on behalf of direct air capture and carbon removal in general. To reinforce the waste management perspective he espouses, he has occasionally handed people plastic bags filled with sand based on the gas mileage of their vehicles—an alternative way of thinking about a colorless, odorless pollutant. The average gas mileage of vehicles on U.S. roads of about twenty-five miles per gallon equates to nearly a one-pound bag of sand for every mile. "Most people were shocked by how much stuff comes out of their car," said Klaus. "People can still be convinced to clean up after themselves regardless of their views on climate change." One thing he's not spending time on is participating in the XPRIZE Carbon Removal competition. He's concerned it would be a distraction from getting direct air capture technology commercialized quickly enough, which, if successful, would negate the need for the XPRIZE proceeds (he also feels burned by its predecessor, the Virgin Earth Challenge).

Instead, he would like to see money for direct air capture committed up front, similarly to the coronavirus vaccine prepurchase agreements that dramatically accelerated the development of vaccines. He suggests using multiple rounds of either public or private procurement, whereby the bid price is dropped in each round until a volume threshold is met. Say you want to get the price of carbon removal below $100 per ton to prove that the technology is commercially feasible. If the learning rate (the rate at which costs decrease with a doubling of output) is assumed to be 20 percent, then the pool of money that should be needed for a given technology and unit size can be estimated. Klaus suggests then boosting that amount, to say half a billion dollars, to allow for more than one technology.[11] "Two things can happen by the time you're out of money," said Klaus. "One is that you got the price down, and that is enormously valuable for society because now you have a tool that can actually solve the problem. Or, you find out that for some obscure reason, this technology refuses to learn." That would run counter to the vast majority of technologies that researchers have looked at—cars, airplanes, computers, mobile phones. "I think the odds are pretty good that it will work. And even if it fails, you learn something too—namely, you learn that you really have to get from Plan A to B, because Plan A is not going to rescue us. Alternatively, you actually manage to solve the problem and it's now below $100 per ton and you can start putting regulations in place to start solving the bigger problem."

That is not to suggest that other carbon removal options won't also contribute to solving the problem—the more options the better—but Klaus sees challenges with relying on any one

---

11 Assuming smaller unit sizes, he suggests that $200 million might be sufficient to drive the cost down to $100 per ton. But that's for one company; more funding would allow for more distinct approaches and learning opportunities. Carbon Engineering's first facility, DAC 1, which will capture over half a million tons per year, will cost more than that itself, but it's expected to start at a lower cost per ton because of its scale and follow a different rate of learning.

nature-based option on its own. So, what is Plan B then? "I am convinced if the crisis really hits, we will do the cheapest and dumbest thing we can do." As it stands today, he believes that would be deploying solar geoengineering without having sufficiently researched its impacts. When his mentor, Wally Broecker, knew he didn't have long to live, he joined forty esteemed climate science colleagues meeting at Arizona State University in 2019 to discuss climate solutions. Unable to join in person, Broecker opened the conference virtually from a wheelchair, breathing through an oxygen tube. He raised concern about the additional surprises that could be expected in the "greenhouse" with continued inaction on stopping the emission of planet-warming gases. "If we are going to prevent the planet from warming up another couple of degrees, we are going to have to go to geoengineering," said Broecker. He had first suggested solar geoengineering be researched in the 1980s, estimating it would take about 700 Boeing 747s to release an amount of sulfur dioxide equivalent to the Mount Pinatubo explosion. A week after addressing the scientists, Wally Broecker died.

During the conference, Klaus raised concerns that ramping up deployment of solar geoengineering too quickly carries the risk of overlooking danger signs that might be more apparent with a more gradual buildup. To do it properly would require a stepwise buildup: deploy, monitor, deploy, monitor… and that takes time. "I'm not sure where the balance is, but I'm a strong advocate that it needs to be researched. It could very well be that solar geoengineering could help us out an awful lot, and that would be a good thing. We may need it even with carbon removal." As a short-term option, it might allow carbon removal enough time to be impactful. As a long-term fix, he worries that it would become unmanageable, like a tourniquet left on for too long that leads to gangrene.

If the ultimate carbon removal objective becomes reducing carbon dioxide's atmospheric concentration by one hundred parts per million, say from 450 to 350, or 500 to 400, that equates to removing roughly 1.5 trillion tons of carbon dioxide.[12] [13] The prototype Klaus has designed with Allen Wright at the Center for Negative Carbon Emissions at Arizona State University, a thirty-foot evolution of various precursors named the MechanicalTree, will have the capacity to capture nearly forty tons per year. For illustrative purposes, if there were over a billion of them, similar to the number of cars on the road worldwide, the total capture capacity would be roughly forty billion tons of carbon dioxide per year. At that rate, it would take about forty years to capture over 1.5 trillion tons of carbon dioxide, or about one hundred parts per million.

Manufacturing has begun, courtesy of $2.5 million in funding from the U.S. Department of Energy that made its way to the Center for Negative Carbon Emissions. Outside the lab sits a concrete pad, surrounded by twelve-foot wrought iron fencing painted in the university's signature maroon tone, which now hosts the first MechanicalTree, manufactured by hand in Wisconsin. The MechanicalTree is a cylindrical tower composed of a central pole that holds 150 carbon dioxide-collecting disks—akin to the leaves on a tree—each exposing the sorbent to air blowing in any direction. After about one hour of exposure, the center pole collapses, lowering the disks mechanically into a regenerative chamber at the base that allows for the captured carbon dioxide to be separated. Then the contraption extends itself again to its full height and the process repeats itself.

---

12 Atmospheric carbon dioxide concentration of 430 parts per million equates to about 1.5 degrees Celsius of warming.

13 Roughly half of carbon dioxide emissions have been absorbed by oceans and biomass, which are in partial equilibrium with the atmosphere. Accordingly, removing carbon dioxide from the atmosphere would trigger outgassing from the oceans and biosphere to restore balance. A removal of one hundred parts per million from the atmosphere, which in isolation would equate to roughly 750 billion tons of carbon dioxide, therefore requires the removal of roughly 1,500 billion tons.

The carbon dioxide from this particular MechanicalTree, which captures carbon dioxide at a rate about a thousand times faster than an actual tree of similar size, will be recirculated to the atmosphere.

After examining the MechanicalTree's performance, and harvesting any learning opportunities, a second one will be constructed and installed adjacent to the university's algae farm, where captured carbon dioxide will fuel growth of the simple, plant-like organisms with numerous potential uses. The third generation will be part of a commercial project being developed by Carbon Collect, a European company that is licensing the technology. "This is the auxons coming through again," said Klaus. Rather than self-replicating colonies of auxons, he envisions mass-produced artificial trees making their way from factories to air capture farms, each hosting as many as 100,000. So long as there is a market for the carbon dioxide.

EPILOGUE

THE U.S. DOE (DEPARTMENT OF ENERGY) organization chart is unwieldy: there are twelve different program offices—among them, the Office of Science, Office of Electricity, Office of Fossil Energy and Carbon Management—all with several sub-offices; and seventeen national research laboratories spread out across the country, each the size of a small town. As their diverse mandates all become increasingly tied to decarbonizing energy, President Joe Biden has tasked them with taking on what have been labeled Earthshots (named in homage to President John F. Kennedy's 1961 moonshot challenge to put a man on the moon within a decade, and not to be confused with the Earthshot Prize launched by Prince William and David Attenborough to award five one-million-pound awards for environmental work annually between 2021 and 2030). The DOE Earthshots aim to overcome the technological and cost hurdles of some of the most consequential decarbonization options, enabling the U.S. to take a leadership role in achieving climate—and competitiveness—objectives. The first Earthshot, the Hydrogen Shot, was announced in June 2021 and sets out to accelerate the innovation and marketability of clean hydrogen, achieving a cost reduction of 80 percent within a decade. The second, the Long Duration Storage Shot, announced a month later, aims to see the cost of longer duration, grid-scale energy storage fall by 90 percent over the same time period. And the third Earthshot, the Carbon Negative Shot, announced in November 2021, will play a meaningful role in reducing the cost of carbon removal, with a goal of getting the price below $100 per metric ton of carbon dioxide equivalent.

To support the Carbon Negative Shot, a new branch was added to the sprawling DOE organization chart in April 2022: CO_2 Removal. It will be temporarily led by Noah Deich, who was seconded from his role leading Carbon180. A primary focus will be helping to implement a $3.5 billion investment in developing four large-scale direct air capture hubs, an initiative that was launched in May 2022 after being included in the $1 trillion infrastructure bill passed at the end of 2021. "Federal funding for carbon removal is at an all-time high, and in my new role I will have an opportunity to direct this support towards a range of efforts that will catalyze innovation, deliver on promises to advance equity and ensure economic benefits and environmental protections for communities," says Noah. Each of the DAC hubs, it's envisioned, will have the capacity to capture and store (or utilize) over one million tons of carbon dioxide per year. The location of existing infrastructure and storage potential will factor in to deciding where to put the hubs, as will the presence of a workforce ready to transition from putting carbon into the sky to removing it. "[The year] 2022 stands to be a significant turning point for the field," says Noah. "Now it is up to the administration and the rest of the DAC ecosystem to ensure that this new funding turns into a tailwind to scale DAC over the coming decades, helping us secure all of the benefits it has to offer."

Around the time Noah joined the DOE's Office of Fossil Energy and Carbon Management, federal legislation was also introduced that would see the DOE become a very serious carbon removal customer. The procurement program functions like the reverse auctions envisioned by Klaus Lackner, just on a larger scale. The proposed annual amount of carbon removal purchased begins at 50,000 tons in 2024 and increases to ten million tons in 2035, while the maximum price for each round decreases from $550 per ton to $150 per ton for a maximum commitment of nearly $10 billion. It would also establish standards for verifying carbon dioxide volumes and aim to galvanize broad public support. "Our bill would help position the U.S. as a world leader in pioneering carbon dioxide removal technology," said Senator Sheldon Whitehouse, a Democrat who co-sponsored the bill and who also delivered 279 speeches on the Senate floor on climate change after the Senate refused to take up a climate bill in 2012. He ceased giving his "Time to Wake Up" weekly speeches when President Biden was elected, saying, "The conditions are at last in place for a real solution . . . it's now time to get to work."

There was insufficient political will to pass the Federal Carbon Dioxide Removal Leadership Act of 2022 as this book was going to print, but other legislation—marking a watershed moment for the climate—was moving ahead. The CHIPS and Science Act, passed on August 9, 2022, is a $280 billion infusion into American ingenuity, seemingly dedicated to revitalizing the U.S. semiconductor industry (CHIPS stands for Creating Helpful Incentives to Produce Semiconductors for America). However, it is also a significant climate bill: roughly one quarter of the funding is to be directed towards accelerating the development of climate-related technologies, including $1 billion for the DOE's Office of Fossil Energy and Carbon Management to spend by 2026 on research and development of carbon removal options. And then, a week later, a landmark in U.S. climate legislation, the Inflation Reduction Act—the name again not entirely reflective of its contents—was enacted. Most significantly for direct air capture, it increases the 45Q tax credit for storing carbon dioxide below ground from $50 per ton to $180 per ton.

The DAC ecosystem is quickly taking shape elsewhere, as well, with a flurry of other announcements coinciding with the U.S. government's skyward leadership push. In June 2022, the DAC Coalition was launched, which aims to build public support for the industry and includes twenty-two technology companies and several other organizations. "Our mission is to serve everyone in the DAC value chain across finance, business, technology, policy, civil society and academia," the coalition stated. "We will also work hand-in-hand with those in the general public, the next generation and frontline communities who have valid questions and concerns." A co-founder of the coalition is Peter Eisenberger's son, Nicholas, a cleantech entrepreneur and investor. He is serving as president of Global Thermostat while a strategic reorientation takes place at the company that includes finding a new CEO and spinning off a nonprofit entity to be led by Graciela Chichilnisky that is tasked with accelerating carbon removal.

Global Termostat's joint development agreement with Exxon was also renewed. "I would say the effort put in to date is fairly immature, so there is a lot of work happening in that space. But that, in my mind, is the holy grail," said Exxon CEO Darren Woods about direct air capture in an interview with CNBC in June 2022. "I think longer term, you move from storing CO_2 to converting it to a product the world needs." (The company also announced that it intends to spend $15 billion over six years on reducing Scope 1 and 2 greenhouse gas emissions that make up 15 percent of the company's total emissions—and twelve times that amount on an annualized basis, $30 billion per year, on share buybacks and dividends.)

Big tech continues to step up as a large commercial buyer of carbon removal. One group, initially made up of Alphabet (Google), McKinsey, Meta (Facebook), Shopify and Stripe, made a commitment in April 2022 to purchase $925 million of carbon removal over nine years. The collaboration, named Frontier, is modeled after vaccine development programs that sent a strong demand signal for a desired outcome without picking winning technologies prematurely. "Frontier is designed to give researchers, entrepreneurs, project developers and investors the confidence to begin building carbon removal technologies today, and to do so with urgency," says Stacy Kauk of Shopify. Another coalition, First Movers, announced shortly thereafter that it had commitments from Alphabet (again), Microsoft and Salesforce to purchase $500 million of carbon removal by 2030.

In the span of several weeks, in the spring of 2022, a significant carbon removal market has been born. Now, instead of being characterized by a lack of demand for carbon removal, the challenge may be finding enough supply until the resounding demand signals entice more participants to enter the race. In the meantime, the leading DAC companies will fill the void as best they can. Climeworks raised $650 million in equity financing to facilitate scaling up to the multimillion-ton capacity. And Carbon Engineering announced that it will be building another facility with up to a million tons of capacity, this time in Scotland. One of the first customers to sign up was Virgin Atlantic.

NOTES

I provide sources below for pertinent details found in the text, both to show a trail of reporting and to provide a resource for those who may be interested in further reading. Publicly spoken words at press conferences, speeches or interviews and the like were taken verbatim and are widely available and as such are not cited here unless deemed worthwhile for providing additional context. Most quotes found in the book are from interviews I conducted. In those cases where I interacted with an individual, whether directly or by email, I have referred to them by their first name (one exception is Graciela Chichilnisky, with whom I only interacted indirectly through Peter Eisenberger).

INTRODUCTION

3   A British science writer named Oliver Morton was also at the meeting: O. Morton, *The Planet Remade: How Geoengineering Could Change the World* (Princeton: Princeton Press, 2015), 27.

CHAPTER 1—FALSE COMPETITION

6   A key influencer was a British scientist named James Lovelock, who had helped others reflect more deeply on humans' relationship with their planet: R. Branson, *Finding My Virginity* (New York: Penguin Random House, 2017), 157.

10   One, *The Weather Makers*, written by an Australian scientist named Tim Flannery, had a profound effect on him: R. Branson, *Losing My Virginity* (New York: Penguin Random House, 2007), 495.

10   Flannery was invited back to the island to talk about the Virgin Earth Challenge: T. Flannery, *An Explorer's Notebook* (New York: HarperCollins, 2007), 268.

11   Gore had voiced misgivings about Lovelock's pessimistic view: R. Branson, *Screw It, Let's Do It* (London: Virgin Books, 2009), 141.

11   Indeed, Lovelock characterized Branson's concept as futile: G. Boynton, "What Makes Richard Run?" *Condé Nast Traveler* (July 14, 2009): https://www.cntraveler.com/stories/2009-07-14/what-makes-richard-run.

11   And hubristic: J. Goodell, "James Lovelock, The Prophet," *Rolling Stone* (November 1, 2007): https://www.rollingstone.com/politics/politics-news/james-lovelock-the-prophet-192646/.

12   In fact, he attributed the book itself to his involvement in the competition: T. Hamilton, "Finding Hope Within the Doom and Gloom of Climate Change,"

Toronto Star (October 28, 2015): https://www.thestar.com/news/world/2015/10/28/finding-hope-within-the-doom-and-gloom-of-climate-change.html.

13 It is believed that he heated potassium nitrate in a metal container: J. Tabor, *Blind Descent: The Quest to Discover the Deepest Place on Earth* (New York: Random House, 2010), 45.

CHAPTER 2—THE ORIGINAL SKY SCRUBBER

17 Klaus and his friend Christopher Wendt co-authored a paper in 1994 laying out their scheme: K. Lackner and C. Wendt, "Exponential Growth of Large Self-Producing Machine Systems," *Permagon Press*, Vol. 21, No. 10, (1995): 55–81.

21 He decided to publish a paper, along with two Los Alamos colleagues: K. Lackner et al., "Carbon Dioxide Extraction from Air: Is It an Option?" Los Alamos National Lab (1999): https://digital.library.unt.edu/ark:/67531/metadc715467/.

23 It wasn't installed at a sufficient scale, though, so eventually oxygen was trucked in: A. Alling et al., *Life Under Glass: Crucial Lessons in Planetary Stewardship from Two Years in Biosphere 2*, Volume 2 (Santa Fe: Synergetic Press, 2020), 169.

24 Bannon tried to find a way to earn some revenue from Biosphere 2: C. Bruck, "How Hollywood Remembers Steve Bannon," *New Yorker* (April 24, 2017): https://www.newyorker.com/magazine/2017/05/01/how-hollywood-remembers-steve-bannon.

24 He responded by threatening to: ibid.

25 After the 2016 U.S. presidential election, Broecker became alarmed when Steve Bannon: K. Krajick, "Wallace Broecker, Prophet of Climate Change," Columbia Climate School (February 19, 2019): https://news.climate.columbia.edu/2019/02/19/wallace-broecker-early-prophet-of-climate-change/.

26 Nothing was said about auxons, but Broecker still thought he was nuts: W. Broecker, "Air Capture of CO_2," *Geochemical Perspectives*, Section 33, p. 332–336 (April 1, 2012): https://pubs.geoscienceworld.org/perspectives/article-abstract/1/2/332/138816/Section-33-Air-Capture-of-CO2.

26 What captured his listeners' attention most was his explanation for how it was economically feasible to scrub the atmosphere of carbon dioxide: W. Broecker and R. Kunzig, *Fixing Climate* (New York: Hill and Wang, 2008), 234.

29 However, his trip went so smoothly in ice-free waters that he had difficulty reconciling his experience: A. Snider and J. Piaskowy, "Comer Conference Scientists Find the Future of Climate Change in Its Past," *Medill Magazine* (October 2, 2009): https://climatechange.medill.northwestern.edu/2015/02/09/comer-conference-scientists-find-the-future-of-climate-change-in-its-past/.

29 "Klaus is the most brilliant person I have ever dealt with": R. Blaustein, "Physicist Turned Carbon-Catcher," *Symmetry Magazine* (December 18, 2014): https://www.symmetrymagazine.org/article/december-2014/physicist-turned-carbon-catcher.

33 He has used an analogy of separating marbles of different colors: H. Herzog, *Carbon Capture* (Cambridge: The MIT Press, 2018), 128–132.

34 To defend his perspective, he has pointed to a plot first developed in the 1950s called the Sherwood Plot: H. Herzog and M. Ranjan, "Feasibility of Air Capture," *Energy Procedia*, Volume 4 (2011): 2871, https://www.sciencedirect.com/ science/article/pii/s1876610211003900.

34 Wally Broecker wrote a research paper that attempted to arbitrate the matter: W. Broecker, "Does Air Capture Constitute a Viable Backstop Against a Bad CO_2 Trip?" *Elementa: Science of the Anthropocene* (2013): https://www8.gsb. columbia.edu/leadership/sites/leadership/files/wally.pdf.

CHAPTER 3—THERMOSTATIC AMBITIONS

37 Instead, Exxon shifted focus to undermining climate change mitigation: A. Powell, "Tracing Big Oil's PR War to Delay Action on Climate Change," *Harvard Gazette* (September 28, 2021): https://news.harvard.edu/gazette/ story/2021/09/oil-companies-discourage-climate-action-study-says/.

37 In practice, it was difficult to shift scientific focus to pressing societal issues: C. Holden, "Earth Institute Director Bows Out," *Science*, Vol. 284, No. 5412 (April 9, 1999): 231–233, https://www.science.org/doi/10.1126/ science.284.5412.231.

39 She believes that for women to be successful professionally: G. Chichilnisky, "Sex and the Ivy League," Columbia University (November 2003): https://papers. ssrn.com/sol3/papers.cfm?abstract_id=1526088.

40 Initially, she focused on the value placed on land resources in the global south: G. Chichilnisky, "North-South Trade, Property Rights, and the Dynamics of the Environment," *International Law and Economics* (1994): https://papers.ssrn. com/sol3/papers.cfm?abstract_id=1377707.

45 An article published in *Bloomberg* said that, according to insiders: L. Kaufman and A. Rathi, "A Carbon-Sucking Startup Has Been Paralyzed by Its CEO," *Bloomberg Green* (April 8, 2021): https://www.bloomberg.com/news/features/ 2021-04-09/inside-america-s-race-to-scale-carbon-capture-technology.

45 The company contracted to build it, Streamline Automation, sued Global Thermostat for $600,000 in unpaid bills, damages and interest: ibid.

CHAPTER 4—HARD SCRUBBING

48 Some of it is sold, then transported to another location and pumped back below the surface to be used for enhanced oil recovery: M. Parker et al., "CO_2 Management at ExxonMobil's LaBarge Field, Wyoming, USA," *Energy Procedia* (2011): 5455–5470, https://www.sciencedirect.com/science/article/pii/ s1876610211008101.

49   A 2017 paper by economists Oliver Hart and Luigi Zingales: O. Hart and L. Zingales, "Companies Should Maximize Shareholder Welfare Not Market Value," *Journal of Law, Finance and Accounting* (2017): 247–274, https://scholar.harvard.edu/files/hart/files/108.00000022-hart-vol2no2-jlfa-0022_002.pdf.

53   Engine No. 1's pitch to entice shareholders to vote for the four nominees at the 2021 annual general meeting: "Reenergize ExxonMobil," Engine No. 1 Presentation (2021): https://assets.contentstack.io/v3/assets/bltc7c628ccc85453af/bltab8cd50fae615bf1/6131f287504fe365615e557d/Engine-No.-1-Reenergize-ExxonMobil-Investor-Presentation.pdf.

57   They spent the break phoning the largest shareholders: J. Camille Aguirre, "The Little Hedge Fund Taking Down Big Oil," *New York Times* (June 23, 2021): https://www.nytimes.com/2021/06/23/magazine/exxon-mobil-engine-no-1-board.html.

58   His rebuttal to Engine No. 1 the only time they did speak directly: ibid.

CHAPTER 5—TWO OPTIONS

61   One natural source is phytoplankton, the typically microscopic plants: O. Morton, *The Planet Remade: How Geoengineering Could Change the World* (Princeton: Princeton Press, 2015), 275.

62   The sulfur dioxide that made its way to the stratosphere created an aerosol mist: S. Self et al., "The Atmospheric Impact of the 1991 Mount Pinatubo Eruption," *Fire and Mud* (1996): https://pubs.usgs.gov/pinatubo/self/#:~:text=Effects%20on%20climate%20were%20an,the%20Earth%20in%201992%2D93.

62   As a result, to what extent solar geoengineering: Morton, *The Planet Remade*, 279.

64   His long-time colleague, Ken Caldeira, has stressed that advocating for solar geoengineering research: J. Anshelm and A. Hansson, "Has the Grand Idea of Geoengineering as Plan B Run Out of Steam?" *Anthropocene Review* (October 29, 2015): https://journals.sagepub.com/doi/10.1177/2053019615614592.

64   Another solar geoengineering researcher who has worked with David, Jason Blackstock: J. Anshelm and A. Hansson, "The Last Chance to Save the Planet? An Analysis of the Geoengineering Advocacy Discourse in the Public Debate," *Environmental Humanities*, Vol. 5 (2014): 101–123, https://www.environmentandsociety.org/mml/last-chance-save-planet-analysis-geoengineering-advocacy-discourse-public-debate.

67   In 1992, they published a paper that separated climate action into three categories: D. Keith and H. Dowlatabadi, "A Serious Look at Geoengineering," *Eos*, Vol. 73, No. 27 (1992): https://keith.seas.harvard.edu/publications/serious-look-geoengineering.

67 A paper he co-authored in 1995: D. Keith and M. Morgan, "Subjective Judgments by Climate Experts," *Environmental Science & Technology*, Vol. 29, No. 10 (1995): https://keith.seas.harvard.edu/publications/subjective-judgments-climate-experts.

67 A century previously, a Swedish scientist named Svante Arrhenius made the first rigorous attempt: A. Lapenis, "Arrhenius and the Intergovernmental Panel on Climate Change," *Eos*, Vol. 79, No. 23 (June 1998): https://www.researchgate.net/publication/248818051_Arrhenius_and_the_intergovernmental_panel_on_climate_change.

68 Arrhenius hypothesized that variations in carbon dioxide levels were likely the cause: E. Crawford, "Arrhenius' 1896 Model of the Greenhouse Effect in Context," *Ambio*, Vol. 26, No. 1 (February 1997): https://www.jstor.org/stable/4314543?seq=1&cid=pdf-reference#references_tab_contents.

76 "Put it this way," he said. "If I had a lump of money to invest to achieve some goal": M. Tamman, "The Adviser," *Reuters* (April 22, 2021): https://www.reuters.com/investigates/special-report/climate-change-scientists-caldeira/.

77 David wrote an op-ed for the *New York Times* in 2021: D. Keith, "What's the Least Bad Way to Cool the Planet," *New York Times* (October 1, 2021): https://www.nytimes.com/2021/10/01/opinion/climate-change-geoengineering.html.

CHAPTER 6—DAC 1

84 "[The Department of Energy] did a hell of a lot of work, and I can't give them enough credit for that": M. Shellenberger, "Interview with Dan Steward, Former Mitchell Energy Vice President," *The Breakthrough Institute* (December 12, 2011): https://thebreakthrough.org/issues/energy/interview-with-dan-steward-former-mitchell-energy-vice-president.

84 Shortly after the creation of the Department of Energy, Schlesinger launched the Carbon Dioxide Effects Research and Assessment Program: "Carbon Dioxide Effects Research and Assessment Program," United States Department of Energy (1980): https://www.osti.gov/servlets/purl/5046001.

84 She studied whatever she could from afar—his mindset, his plays, his leadership tips: M. Swartz, "How the Most Hyped U.S. Oil Merger in a Decade Went Bust," *Texas Monthly* (February 2021): https://www.texasmonthly.com/news-politics/how-the-most-hyped-u-s-oil-merger-in-a-decade-went-bust/.

89 His holding company, Berkshire Hathaway, purchased another large stake in the company, about $7 billion in common shares: A. Crippen, "Warren Buffett Scoops Up Another $1 Billion in Occidental Shares, Bringing Total Stake to $7 Billion," CNBC (March 17, 2022): https://www.cnbc.com/2022/03/17/warren-buffett-scoops-up-another-1-billion-in-occidental-shares-bringing-total-stake-to-7-billion.html.

CHAPTER 7—ORCA

94 Of more immediate concern as Jan and Christoph sought to ramp up their business was a 2011 report from the American Physical Society: "Direct Air Capture of CO_2 with Chemicals," American Physical Society (2011): https://www.aps.org/policy/reports/assessments/dac-biblio.cfm.

96 The extra carbon dioxide accelerates the photosynthesizing of vegetables: E. Kolbert, *Under a White Sky: The Nature of the Future* (New York: Crown, 2021), 157.

96 The flipside is that scientists have found that food grown at elevated carbon dioxide levels is less nutritious: A. Sneed, "Ask the Experts: Does Rising CO_2 Benefit Plants?" *Scientific American* (January 23, 2018): https://www.scientificamerican.com/article/ask-the-experts-does-rising-co2-benefit-plants1/.

97 Klaus had written a paper in 1995 with colleagues from Los Alamos: K. Lackner et al., "Carbon Dioxide Disposal in Carbonate Minerals," Los Alamos National Laboratory (1995): https://www.sciencedirect.com/science/article/abs/pii/036054429500071N.

97 Typically, carbon takes a long time to make its way through the slow geological cycle: H. Riebeek, "The Carbon Cycle," NASA Earth Observatory (June 16, 2011): https://earthobservatory.nasa.gov/features/CarbonCycle.

100 It took less than a year for the basalt to absorb over 95 percent of the injected carbon dioxide: S. Gíslason et al., "A Brief History of CarbFix: Challenges and Victories of the Project's Pilot Phase," *Energy Procedia*, Vol. 146 (July 2018): 103-114, https://www.sciencedirect.com/science/article/pii/S1876610218301462.

100 Furthermore, the cost of drilling, injecting and monitoring was estimated at only about $5 per ton: I. Gunnarsson, "The Rapid and Cost-Effective Capture and Subsurface Mineral Storage of Carbon and Sulfur at the CarbFix2 Site," *International Journal of Greenhouse Gas Control*, Vol. 79 (December 2018): 21, https://eprints.soton.ac.uk/425577/1/Gunnarsson_et_al._IJGGC_manuscript_revised.pdf.

101 The storage capacity of Iceland's basaltic rocks is estimated to be over 2.5 trillion tons: ibid., 25.

CHAPTER 8—SOLAR-POWERED PATHWAY

104 One of the most widely cited studies looking at cost estimates for proposed environmental regulations: H. Hodges, "Falling Prices: Cost of Complying with Environmental Regulations Almost Always Less than Advertised," *Economic Policy Institute* (November 1997).

104 Acid rain compliance costs in the U.S.: C. Van Atten and L. Hoffman-Andrews, "The Clean Air Act's Economic Benefits: Past, Present and Future," Small Business Majority and The Main Street Alliance (October 2010):

https://smallbusinessmajority.org/sites/default/files/research-reports/Benefits_ of_CAA_100410.pdf.

104 Estimates for controlling volatile organic compounds: ibid.

104 Estimates for controlling air pollution from coke ovens: H. Hodges, "Falling Prices: Cost of Complying with Environmental Regulations Almost Always Less than Advertised," *Economic Policy Institute* (November 1997).

105 And sulfur dioxide regulations imposed on electric utilities: ibid.

105 In 2008, David published a paper with two colleagues that advised governments: D. Keith et al., "Expert Assessments of Future Photovoltaic Technologies," *Environmental Science & Technology*, Vol. 42, No. 24 (2008): https://keith.seas.harvard.edu/publications/ expert-assessments-future-photovoltaic-technologies.

105 A few years later, Greg, now a professor at University of Wisconsin-Madison, completed a survey along with colleagues of sixty-five solar experts: G. Nemet, *How Solar Became Cheap* (New York: Routledge, 2019), 11-12.

109 The country is now responsible for nearly two-thirds of worldwide solar photovoltaic manufacturing: "Solar PV Module Shipments by Country of Origin, 2012-2019," IEA (May 26, 2020): https://www.iea.org/data-and-statistics/ charts/solar-pv-module-shipments-by-country-of-origin-2012-2019.

109 The group set an efficiency record for solar photovoltaics: R. Kurmelovs, "Insanely Cheap Energy: How Solar Power Continues to Shock the World," *Guardian* (April 21, 2021): https://www.theguardian.com/australia-news/2021/apr/25/ insanely-cheap-energy-how-solar-power-continues-to-shock-the-world.

113 In the case of COVID-19 vaccines, nearly $100 billion worth of these agreements were signed in 2020: "Governments Spent at Least €93bn on COVID-19 Vaccines and Therapeutics During the Last 11 Months," *Business Wire* (January 11, 2021): https://www.businesswire.com/news/home/20210110005098/en.

115 A litany of factors have been blamed by the company developing it: S. Kelly, "How America's Clean Coal Dream Unraveled," *Guardian* (March 2, 2018): https://www.theguardian.com/environment/2018/mar/02/clean-coal- america-kemper-power-plant.

CHAPTER 9—CUSTOMER FEEDBACK

120 "We've bought forest offsets that are now burning": C. Hodgson,"US Forest Fires Threaten Carbon Offsets as Company-Linked Trees Burn," *Financial Times* (August 2, 2021): https://www.ft.com/content/3f89c759-eb9a-4dfb-b768- d4af1ec5aa23.

122 An aerospace engineer named Kevin Meissner was looking for impactful problems: E. Pontecorvo, "Meet the Startup Producing Oil to Fight Climate Change," *Grist* (May 18, 2021): https://grist.org/climate-energy/lucky-charm/.

123 Tim Flannery, the Australian scientist and author, has investigated many of them: T. Flannery, "Removing 10 Gigatons of Carbon Dioxide," Presentation to AirMiners (September 3, 2021): https://www.youtube.com/watch?v=SRVnitJIr2c.

125 As a result, one type of seaweed, giant kelp, can grow up to two feet per day: T. Flannery, *Sunlight and Seaweed: An Argument for How to Feed, Power and Clean Up the World* (Text Publishing, 2017), 59–61.

125 It was envisioned that a 100,000-acre farm could be built at a cost of $2–$4 billion: E. Holles, "Scientists Begin Effort to Produce Methane Gas and Nutrients from Kelp Off Southern California Coast," *New York Times* (July 13, 1976): https://www.nytimes.com/1976/07/13/archives/scientists-begin-effort-to-produce-methane-gas-and-nutrients-from.html.

126 Over a twelve-year period, about $20 million was spent on the offshore experiments: J. Kim et al., "Opportunities, Challenges and Future Directions of Open-Water Seaweed Aquaculture in the United States," *Phycologia*, Vol. 58, No. 5 (2019): 446–461, https://www2.whoi.edu/site/lindell-lab/wp-content/uploads/sites/91/2020/04/Opportunities-challenges-and-future-directions-of-open-water-seaweed-aquaculture-in-the-United-States-1.pdf.

131 Overall, Lucas remains optimistic that progress will indeed be made: T. Raftery, "Microsoft's Incredibly Ambitious Climate Emissions Initiative—A Chat with Lucas Joppa," *Climate 21*, Season 1, Episode 4 (December 16, 2020).

CHAPTER 10—HYPERION II

132 Writing about the experience: A. Huxley, *Adonis and the Alphabet and Other Essays* (Chatto and Windus, 1956).

133 The courts were supportive, typically deciding in favor of polluting businesses: J. Benidickson, *The Culture of Flushing: A Social and Legal History of Sewage* (Vancouver: UBC Press, 2007), 255, 284–285.

135 The Cuyahoga had erupted in flames before: J. Latson, "The Burning River That Sparked a Revolution," *Time* (June 22, 2015): https://time.com/3921976/cuyahoga-fire/.

136 Its absence was discovered by a high school teacher named Howard Bennett: S. Randle, "The Effluent Society," *The Awl* (March 25, 2016): https://www.theawl.com/2016/03/the-effluent-society.

136 In arguing their position, city officials largely relied on testimony from a former mining engineer named William Bascom: A. Sklar, *Brown Acres* (Santa Monica: Angel City Press, 2008), 183, 186, 196, 205.

138 Along with publishing their findings: J. Shapiro et al., "The Low but Uncertain Measured Benefits of US Water Quality Policy," *Proceedings of*

the National Academy of Sciences (March 19, 2019): https://www.pnas.org/content/116/12/5262#ref-43.

139 In looking at the 1990 Clean Air Act amendments: "The Benefits and Costs of the Clean Air Act from 1990 to 2020," Environmental Protection Agency (2011): https://www.epa.gov/sites/default/files/2015-07/documents/fullreport_rev_a.pdf.

139 Another study, which applied a willingness-to-pay approach to simply having clean water, estimated the cost at $4.9 billion: C. Kling et al., "From Exxon to BP: Has Some Number Become Better than No Number?" *Journal of Economic Perspectives*, Vol. 26, No. 4 (2012): https://www.aeaweb.org/articles?id=10.1257/jep.26.4.3.

140 How much the beaches will be impacted depends on how much warming occurs: J. Aerts et al., "Pathways to Resilience: Adapting to Sea Level Rise in Los Angeles," *Annals of the New York Academy of Sciences*, Vol. 1427 (September 2018): https://nyaspubs.onlinelibrary.wiley.com/toc/17496632/2018/1427/1.

141 If the temperature increase is limited to 1.5 degrees, however, sea level rise by 2300 could be held below two feet: "The Impacts of Climate Change at 1.5C, 2C and Beyond," *CarbonBrief* (accessed in 2021): https://interactive.carbonbrief.org/impacts-climate-change-one-point-five-degrees-two-degrees/?utm_source=WEB&utm_campaign=Redirect#.

142 To model L.A.'s transition, an unprecedented effort was undertaken: S. Roth, "L.A. Wants to Ditch Fossil Fuels, Researchers Say It's Possible," *Los Angeles Times* (March 24, 2021): https://www.govtech.com/fs/la-wants-to-ditch-fossil-fuels-researchers-say-its-possible.html.

142 An investigation by the *Los Angeles Times*, *Floodlight* and the *Guardian*: M. Green, "They Fought for Clean Air. They Didn't Know They Were Part of a Gas Industry Campaign," *Guardian* (August 11, 2021): https://www.theguardian.com/environment/2021/aug/11/los-angeles-long-beach-natural-gas-trucks.

143 A couple of years later it emerged that SoCalGas had paid consultants: "Public Advocates Office Investigation into SoCalGas Pro-Gas Advocacy," Public Advocates Office, California Public Utilities Commission (accessed in 2022): https://www.publicadvocates.cpuc.ca.gov/general.aspx?id=4294.

143 The entire amount available in the U.S. is equivalent to about a quarter of the amount of gas used by California's buildings alone: D. Roberts, "The False Promise of 'Renewable Natural Gas,'" *Vox* (February 20, 2020): https://www.vox.com/energy-and-environment/2020/2/14/21131109/california-natural-gas-renewable-socalgas.

144 A 2021 study by researchers at Cornell and Stanford universities: R. Howarth and M. Jacobson, "How Green Is Blue Hydrogen?" *Energy Science and Engineering* (August 12, 2021): https://doi.org/10.1002/ese3.956.

CHAPTER 11—THE PRIZE

148 A greater sustainability concern might be black carbon (or soot): R. Ryan et al., "Impact of Rocket Launch and Space Debris Air Pollutant Emissions on Stratospheric Ozone and Global Climate," *Earth's Future* (June 9, 2022): https://agupubs.onlinelibrary.wiley.com/doi/10.1029/2021EF002612.

158 He suggests using multiple rounds of either public or private procurement: K. Lackner and H. Azarabadi, "Buying Down the Cost of Direct Air Capture," *Industrial and Engineering Chemical Research*, Vol. 60, No. 22 (May 26, 2021): https://pubs.acs.org/doi/10.1021/acs.iecr.0c04839.

159 He had first suggested solar engineering be researched in the 1980s: W. Stevens, "Scientist at Work: Wallace S. Broecker; Iconoclastic Guru of the Climate Debate," *New York Times* (March 17, 1998): https://www.nytimes.com/1998/03/17/science/scientist-at-work-wallace-s-broecker-iconoclastic-guru-of-the-climate-debate.html.

ACKNOWLEDGMENTS

WRITING THIS BOOK was an awe-inspiring experience; I am very grateful to have had the opportunity to tell the story of the remarkable people in it, and for their assistance in doing so. It wouldn't have been possible without Ed Whittingham; in addition to being a remarkable person himself who tirelessly advocates for carbon removal, Ed played an instrumental role in convincing the Canadian government to propose a 60 percent tax credit for direct air capture that might be passed by the time this book is printed. Ed was pivotal in introducing me to key individuals, and he collaborated on research and provided sage advice along the way.

Thank you to Figure 1 for being a great publishing partner, especially Steve Cameron, Melissa Churchill and David Marsh for their editing prowess and Teresa Bubela, Mark Redmayne and Lara Smith for contributing further polish and support. Valuable contributions were also made by Adam Bink, Stephen Bown, John Cirucci, Erica Dodds, Kevin Fitzgerald, Jackson Hegland, Travis Johnson, Jason Kmon, Bob Page, Denise Rowe, Bodhi Thakur and Eva Voinigescu.

Finally, I am grateful to friends and family for their support. In particular, I'd like to thank my parents, Patti and Bob McKendrick, who continue to be a devoted support crew and find ways to be helpful with many elements of the book writing process. And to my wife, Kylie, who graciously minded the shop while I was absent—and absentminded—and is a source of inspiration.

INDEX

Lands' End, 29
land-use issues, 102, 123, 127
learning rate effect, 113-15. *See also*
scaling up
Lehman Brothers, 35
limestone, 32, 98, 124
lobbying, 115, 134, 135-36, 142-43
Long Duration Storage Shot, 163
Longitude (Sobel), 9
Los Alamos National Laboratory, 18,
25, 97
Los Angeles, California, 132-35, 137,
140, 141-42, 144-46. *See also*
California
Los Angeles Times, 142, 145
Louisiana, 89
Lovelock, James, 6-7, 11, 61
low-carbon fuels, 87
Low Carbon Solutions, 56, 59
Lütke, Tobi, 128

machine learning, 118
Mann, Michael, 64
marine life, 136
market demand, 14, 112, 116, 161, 165
Mars, 6, 7, 61, 147, 148
Massachusetts Insitute of Technology
(MIT), 22, 38-39, 137-38
mathematics, 28
Matter, Juerg, 98-99
McDermott, Jim, 78-79, 83, 87, 89,
91, 145-46
McKinsey, 165
MechanicalTree, 160-61
Meehan, Shaun, 122-23
Meissner, Kevin, 122
Meta, 165
metal oxides, 16
methane, 135, 144
Microsoft, 69-70, 76, 117-21, 119n7,
123, 127, 130-31, 165
mineralization, 25, 98-99, 102
Mississippi, 115
Mitchell, George, 83-84
Mitchell Energy, 84
moisture swing phenomenon, 33
moral hazards, 75, 117
Moreton, Oliver, 3-4
Morocco, 96

Mount Pinatubo, 61-62
Munger, Charlie, 88
municipal bonds, 145
Musk, Elon, 147-48, 151, 152
Musk Foundation, 148

National Petroleum Council, 52
National Renewable Energy
Laboratory, 142, 143
natural gas, 49, 142-44
nature-based approaches, 12, 120, 127
Necker Island, 9, 10
Nemet, Greg, 105-7, 109, 111-14, 116
New Yorker, 26
New York Times, 77
Night of the Long Batons, 38
Nixon, Richard, 136
Nortel Networks, 149
Northwest Passage, 29
NRG Energy, 82, 150, 152
nuclear power, 90-91, 113-14

Occidental Petroleum, 84-85, 87, 88
Office of Energy Efficiency and
Renewable Energy, 52
Office of Fossil Energy and Carbon
Management, 163
oil production, 36-37
oil recovery, 85-88
oil spills, 135-36, 139
olivine mineral, 12, 124
Oman, 102
One Two Three... Infinity (Gamow), 17
optimism, 131
Orca, 101-2, 153

Pacific Northwest National
Laboratory, 144
paper production, 72
Penner, Charlie, 51-53, 56-58
peridotite, 102
Permian Basin, 84-85, 88, 89
Perón, Eva, 37-38
Perón, Juan, 38
Petronas, 55
photosynthesis, 36, 41
photosynthesis, artificial, 44-45
physical chemistry, 67
physics, 17-18, 67, 71, 149-50

SCRUBBING THE SKY: THE PODCAST

Join Ed Whittingham and Paul McKendrick as they take listeners on an eight-part audio tour of the evolution of direct air capture's use to scrub the sky of carbon dioxide. Over the course of the series, Ed and Paul explore the birth of this important new industry, drawing on interviews to tell the stories of the researchers, developers, CEOs, funders, and politicians behind it.

Scrubbing the Sky: The Podcast can be accessed at **scrubbingthesky.com**.

ABOUT THE HOSTS

ED WHITTINGHAM is a co-host of *Energy vs. Climate*, a leading podcast that explores the tradeoffs and realities of the energy transition from fossil fuels to renewables. He is the former executive director of the Pembina Institute, a national think tank supporting Canada's clean energy transition, and has worked to advance climate and energy policy—including for carbon removal—for two decades. Ed is also a highly experienced public speaker, and his op-eds have been published in newspapers and magazines across Canada and internationally.

Best-selling author **PAUL MCKENDRICK** previously partnered in a firm that developed and financed renewable energy projects, and he has also worked in the electric utility sector, the oil and gas sector, and investment banking. *Scrubbing the Sky* is his second book.